Saida Laalioui

Perovskite-Based Solar Cells

Also of interest

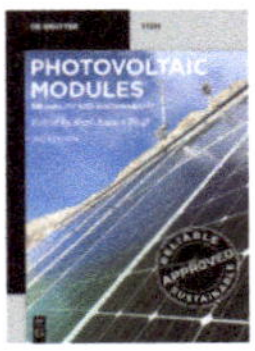

Photovoltaic Modules.
Reliability and Sustainability
Edited by: Karl-Anders Weiß, 2021
ISBN 978-3-11-068554-1, e-ISBN (PDF) 978-3-11-068572-5
e-ISBN (EPUB) 978-3-11-068555-8

Energy and Sustainable Development
Quinta Nwanosike Warren, 2020
ISBN 978-1-5015-1973-4, e-ISBN (PDF) 978-1-5015-1977-2
e-ISBN (EPUB) 978-1-5015-1328-2

Photovoltaic Module Technology
Harry Wirth, 2020
ISBN 978-3-11-067697-6, e-ISBN (PDF) 978-3-11-067701-0
e-ISBN (EPUB) 978-3-11-067710-2

Solar Photovoltaic Generation
Jinhuan Yang, Xiao Yuan and Liang Ji, 2020
ISBN 978-3-11-053138-1, e-ISBN (PDF) 978-3-11-052483-3
e-ISBN (EPUB) 978-3-11-052542-7

Chemistry and Energy.
From Conventional to Renewable
Mark Anthony Benvenuto, 2022
ISBN 978-3-11-066226-9, e-ISBN (PDF) 978-3-11-066227-6
e-ISBN (EPUB) 978-3-11-066233-7

GREEN – Alternative Energy Resources
ISSN 2509-7237, e-ISSN 2509-7245

Saida Laalioui

Perovskite-Based Solar Cells

From Fundamentals to Tandem Devices

DE GRUYTER

Author
Saida Laalioui
Green Energy Park – (IRESEN, UM6P)
R206 Route Régionale
Ben Guerir
Morocco
laalioui@greenenergypark.ma
laalioui@iresen.org

ISBN 978-3-11-076060-6
e-ISBN (PDF) 978-3-11-076061-3
e-ISBN (EPUB) 978-3-11-076065-1

Library of Congress Control Number: 2021947446

Bibliographic information published by the Deutsche Nationalbibliothek
The Deutsche Nationalbibliothek lists this publication in the Deutsche Nationalbibliografie; detailed bibliographic data are available on the Internet at http://dnb.dnb.de.

Typesetting: Integra Software Services Pvt. Ltd.
Printing and binding: CPI books GmbH, Leck

www.degruyter.com

Acknowledgments

First of all, I express my eternal gratitude to the Green Energy Park, a research center that allowed me to flourish scientifically and personally. A research center lead by its great co- President **Mr. Badr IKKEN**, its Manager **Mr. Zakaria Naimi** and its coordinator **Mr. Abdellatif Ghennioui**, three great men that are the cornerstone that continuously encouraged and unconditionally supported the completion of this project, without their trust and precious advice, this book would have remained a simple unconcreted idea.

I also thank my awesome colleague and friend **Kawtar belrhiti Alaoui** for helping me since the concept to the manuscript. Indeed, she was her idea to valuate my research to a book and she did not save any moment to support me from the very first draft.

Last but not least, I am eternally grateful to my Mentor and Professor **Pr. Abdelkader Outzourhit** at Faculty of Science Semlalia- Cadi Ayyad University, a man of science, who brings together wisdom and humility, with an outstanding sense of support, for all the time he devoted, despite his busy schedule to guide my steps, correct, comment and review this manuscript in all its stages. I thank him sincerely for his professionalism, his availability, and his advice.

Concretizing this book was not as easy as I thought at the beginning. So many thanks to all those who supported me, encouraged me and gave their precious time during the different stages of this experience, which was special and unique.

https://doi.org/10.1515/9783110760613-202

Contents

Acknowledgments —— V

Abbreviations list —— XI

List of figures —— XIII

List of tables —— XV

General introduction —— 1

Chapter 1
Perovskite-based solar cells —— 2

1 Introduction —— **2**
2 Perovskites —— **2**
3 Crystal structure of the hybrid perovskite —— **2**
4 Characteristics of perovskites —— **3**
5 Structure of solar cells based on perovskites —— **5**
5.1 Evolution of the structure of perovskite-based solar cells —— **5**
5.2 Description of the structure of the perovskite solar cell —— **5**
5.3 The components of the perovskite solar cell —— **6**
5.3.1 Absorbers —— **6**
5.3.2 Electron transporting layer —— **9**
5.3.3 Hole transporting layer —— **11**
5.3.4 Metal contacts —— **13**
5.3.5 Substrates —— **13**
5.4 Structures of the perovskite solar cells —— **14**
5.4.1 Inorganic n-type/perovskite/organic p-type —— **14**
5.4.2 Organic n-type/perovskite/inorganic p-type —— **14**
5.4.3 Organic n-type/perovskite/organic p-type —— **14**
5.4.4 Inorganic n-type/perovskite/inorganic p-type —— **15**
6 Operating principle of a perovskite solar cell —— **15**
7 The evolution of the perovskite solar cell efficiency —— **15**
8 Conclusion —— **16**

Chapter 2
Perovskite solar cell fabrication methods —— 17

1 Introduction —— **17**
2 One-step spin-coating method —— **17**
3 The hot-casting technique —— **18**
4 The anti-solvent method —— **18**

5 The two-step spin-coating technique — 19
6 The dual-source vapor deposition technique — 19
7 The sequential dual-source vapor deposition technique — 20
8 The adduct method — 21
9 The vapor-assisted solution process (VASP) — 21
10 Ink-jet printing technique — 22
11 Techniques to improve the performance of perovskite — 23
11.1 Techniques to improve perovskite efficiency — 23
11.2 Techniques for improving the perovskite stability — 24
12 Conclusion — 28

Chapter 3
Perovskite-based tandem solar cells — 29
1 Introduction — 29
2 The definition and principle of tandem solar cells — 29
3 Tandem solar cell configurations — 29
3.1 Four-terminal tandem — 29
3.2 Monolithic tandem solar cell — 31
4 Tandem solar cells containing the perovskite sub-cells — 32
4.1 The perovskite/perovskite tandem solar cells — 32
4.1.1 Structure of the perovskite top cell — 32
4.1.2 Structure of the perovskite bottom cell — 33
4.1.3 Tandem solar cell configuration — 33
4.1.4 Characteristics of tandem solar cells and sub-cells — 34
4.1.5 Other perovskite/perovskite tandem solar cells — 35
4.2 Perovskite/CIGS tandem solar cells — 36
4.2.1 Introduction — 36
4.2.2 The perovskite sub-cell — 36
4.2.3 The CIGS sub-cell — 39
4.2.4 The tandem solar cell — 41
4.2.5 Other perovskite/CIGS tandem solar cells — 43
5 Conclusion — 43

Chapter 4
Perovskite/silicon heterojunction tandem solar cells — 44
1 Introduction — 44
2 The silicon heterojunction solar cell — 44
2.1 Description of the silicon heterojunction solar cell — 44
2.2 Evolution of the performances of silicon heterojunction solar cells — 45
2.3 The configuration of the record silicon heterojunction solar cell — 45

2.4 Fabrication method of the record silicon heterojunction solar cell — **46**
3 Principle of the tandem perovskite/silicon heterojunction solar cell — **47**
4 Perovskite/silicon heterojunction tandem solar cell — **47**
4.1 The tandem solar cell of 18.1% efficiency — **47**
4.1.1 The perovskite top cell — **48**
4.1.2 The silicon heterojunction bottom cell — **49**
4.1.3 Monolithic tandem solar cell — **50**
4.2 The tandem solar cell of 21.2% efficiency — **52**
4.2.1 The perovskite top cell — **52**
4.2.2 Silicon heterojunction bottom cell — **55**
4.2.3 The monolithic tandem solar cell — **56**
4.3 The tandem solar cell of 23.6% efficiency — **58**
4.3.1 The perovskite sub-cell — **58**
4.3.2 Silicon heterojunction bottom cell — **61**
4.3.3 The monolithic tandem solar cell — **62**
4.4 Recent progress in perovskite/silicon tandem solar cells — **62**
4.5 Perovskite/silicon monolithic tandem solar cells evolution — **64**
5 Means to improve the performances of the tandem solar cells — **65**
6 Conclusion — **68**

General conclusion — 69

References — 71

Index — 77

Abbreviations list

µc-Si:H	*Hydrogenated microcrystalline silicon*
2D	*Two-dimensional*
3D	*Three-dimensional*
Ag	*Silver*
ALD	*Atomic layer deposition technique*
AR	*Antireflective*
a-Si:H	*Hydrogenated amorphous silicon*
Au	*Gold*
Au-NPs	*Gold nanoparticles*
BCP	*Bathcuproine*
Br	*Bromine*
C60	*Fullerene*
$CH_3NH_3PbI_3$	*Methylammonium lead iodide*
CIGS	*Copper indium gallium di-selenide*
CIS	*Copper indium di-selenide*
Cs	*Cesium*
c-Si	*Crystalline silicon*
CsI	*Cesium iodide*
c-TiO_2	*Compact titanium dioxide*
CuSCN	*Copper thiocyanate*
CVD	*Chemical vapor deposition*
DEPOT	*Poly(3,4-ethylenedioxythiophene)*
DIO	*1,8-Diiodooctane*
DMF	*Dimethylformamide*
DMSO	*Dimethylsulfoxide*
EA	*Ethylammonium*
EPFL	*Federal Institute of Technology in Lausanne*
EQE	*External quantum efficiency*
ETL	*Electron transporting layer*
FA	*Formamidinium*
FAI	*Formamidinium iodide*
FF	*Fill factor*
FTO	*Fluorine-doped Tin Oxide*
HIT	*Heterojunction solar cell*
HTL	*Hole transporting layer*
IO:H	*Hydrogenated indium oxide*
IPA	*2-Propanol*
ITO	*Indium tin oxide*
IZO	*Indium zinc oxide*
J_{sc}	*Short-circuit current density*
J–V	*Current density versus voltage measurements*
LIF	*Lithium fluoride*
MA	*Methylammonium*
MAI	*Methyl ammonium iodide*
$MAPbI_3$	*Methylammonium lead iodide*
$MAPbX_3$	*Methylammonium lead halide*

https://doi.org/10.1515/9783110760613-204

$MASnX_3$	*Methylammonium tin halide*
Mo	*Molybdenum*
MPP	*Maximum power point tracking*
NiO	*Nickel oxide*
NREL	*National Renewable Energy Laboratory*
P	*Pressure*
P3HT	*Poly(3-hexylthiophene-2,5-diyl)*
Pb	*Lead*
$PbBr_2$	*Lead bromide*
PbI_2	*Lead iodide*
PCBM	*Methyl[6,6]-phenyl-C61-butanoate*
PECVD	*Plasma-enhanced chemical vapor deposition*
PEIE	*Polyethylenimine*
PEN	*Polyethylenenaphthalate*
PSCs	*Perovskite solar cells*
PSS	*Poly(styrenesulfonate)*
PTAA	*Poly(triaryl amine)*
SEM	*Scanning electron microscope*
Sn	*Tin*
SnO_2	*Tin dioxide*
Spiro-OMeTAD	*2,2′7,7′-tetrakis(N, N-di-p-methoxyphenylamine)-9,9′-spirobifluorene*
SPO	*Stabilized power output*
T	*Temperature*
TCO	*Transparent conductor oxide*
UV	*Ultraviolet*
V	*Voltage*
V_{oc}	*Open-circuit voltage*
ZnO:Al	*Aluminum-doped zinc oxide*
ZTO	*Zinc tin oxide*

List of figures

Figure 1 Crystalline structure of the hybrid perovskite —— **2**
Figure 2 The absorption spectrum of the perovskite $CH_3NH_3Pb_{(1-X)}Sn_XI_3$. —— **4**
Figure 3 Optical bandgap for the perovskite $CH_3NH_3SnI_{3-x}Br_x$. —— **4**
Figure 4 Evolution of the perovskite-based solar cell structure —— **6**
Figure 5 Standard structure of perovskite solar cell —— **6**
Figure 6 The role of the ETL on the movement of the electrons and holes —— **9**
Figure 7 Perovskite solar cell structure containing AZO as TCO and electron transporting layer —— **10**
Figure 8 Perovskite solar cell structures (a) n-i-p and (b) p-i-n configurations, respectively —— **14**
Figure 9 Operating principle of a perovskite solar cell —— **15**
Figure 10 The evolution of the perovskite-based solar cells —— **16**
Figure 11 One-step spin-coating method —— **17**
Figure 12 The hot-casting technique —— **18**
Figure 13 The anti-solvent method —— **18**
Figure 14 The two-step spin-coating technique —— **19**
Figure 15 The dual-source vapor deposition technique —— **20**
Figure 16 The sequential dual-source vapor deposition technique —— **20**
Figure 17 The adduct method —— **21**
Figure 18 The vapor-assisted solution process (VASP) —— **22**
Figure 19 Ink-jet printing illustration —— **22**
Figure 20 The images (a, b, c, d) correspond to the perovskite films deposited on FTO /compact-TiO_2 substrates from a precursor solution containing amounts of 0, 0.25, 0.5, and 1.0 M acetic acid, respectively —— **23**
Figure 21 The blue curve represents the evolution of the perovskite solar cell efficiency as a function of time, tested in short circuit while the red one is tested in open circuit —— **24**
Figure 22 The size of grains as a function of the temperature of the substrate with the same precursor solution at 70 °C —— **25**
Figure 23 Effect of cesium on the stability of perovskite —— **25**
Figure 24 Effect of mesoporous TiO_2 and SnO_2, and planar SnO_2 on the stability of the perovskite solar cell —— **26**
Figure 25 Effect of the encapsulation on the stability of the perovskite solar cell —— **27**
Figure 26 The effect of the encapsulation and UV filter on the perovskite solar cell degradation —— **27**
Figure 27 Image of perovskite/silicon heterojunction tandem solar cells reaching efficiencies up to 21.2% and 19.2% for cell areas of 0.17 and 1.22 cm^2, respectively —— **30**
Figure 28 Principle of the tandem solar cell —— **30**
Figure 29 Scheme of the four-terminal tandem solar cell —— **31**
Figure 30 Scheme of the monolithic tandem solar cell —— **31**
Figure 31 Structure of the perovskite top cell for the four-terminal and the monolithic tandem solar cells, respectively —— **33**
Figure 32 Structure of the perovskite bottom cell —— **33**
Figure 33 Scanning electron microscopic image of the monolithic perovskite/perovskite tandem solar cell —— **34**
Figure 34 Four-terminal perovskite/perovskite tandem solar cell —— **34**

https://doi.org/10.1515/9783110760613-205

Figure 35 The external quantum efficiency of the two tandem solar cells for (a) monolithic configuration and (b) four-terminal configuration — **35**
Figure 36 A transverse view of the perovskite solar cell by scanning electron microscope (SEM) — **37**
Figure 37 Fabrication method of the perovskite sub-cell — **38**
Figure 38 EQE, total transmission T, and reflection R of the perovskite top cell as a function of wavelength — **39**
Figure 39 The configuration of the CIGS sub-cell — **39**
Figure 40 Method of elaboration of the CIGS sub-cell — **40**
Figure 41 The four-terminal perovskite/CIGS tandem solar cell — **41**
Figure 42 The external quantum efficiency of the sub-cells as well as the tandem solar cell — **42**
Figure 43 J–V measurements of the two sub-cells — **42**
Figure 44 Structure of the silicon heterojunction cell — **45**
Figure 45 The evolution of the efficiency and the current density of the HIT solar cells — **45**
Figure 46 Structure of the record heterojunction solar cell — **46**
Figure 47 PECVD deposition chamber schemes — **46**
Figure 48 Principle of the tandem perovskite/silicon heterojunction solar cell — **47**
Figure 49 A cross section of the monolithic tandem solar cell seen by SEM — **48**
Figure 50 The perovskite top cell structure — **48**
Figure 51 Fabrication method of the perovskite top cell — **49**
Figure 52 The silicon heterojunction bottom cell structure — **49**
Figure 53 Fabrication method of the silicon heterojunction bottom cell — **50**
Figure 54 Monolithic tandem solar cell structure — **51**
Figure 55 SEM cross section of the perovskite top cell illuminated from MoO_x/IO:H/ITO — **52**
Figure 56 Fabrication method of the perovskite top cell — **53**
Figure 57 Performances of perovskite solar cell used in single superstrate junction (illuminated from PCBM) and that used in tandem substrate solar cell (illuminated from spiro-OMeTAD) — **55**
Figure 58 Silicon heterojunction bottom cell structure — **55**
Figure 59 Fabrication method of the silicon heterojunction bottom cell — **56**
Figure 60 Monolithic perovskite/silicon tandem solar cell structure — **57**
Figure 61 *J–V* measurements for the same tandem solar cell with an aperture area of 0.17 cm^2 with and without antireflective layer — **58**
Figure 62 The perovskite sub-cell structure — **59**
Figure 63 Fabrication method of the perovskite top cell — **60**
Figure 64 Silicon heterojunction bottom cell structure — **60**
Figure 65 Fabrication method of the silicon heterojunction bottom cell — **61**
Figure 66 Structure of the monolithic record tandem solar cell — **62**
Figure 67 J–V measurements of the perovskite/silicon tandem solar cell — **63**
Figure 68 Structure of the monolithic tandem solar cell of 25.4% efficiency — **63**
Figure 69 Structure of the monolithic tandem solar cell of 27.13% efficiency — **64**
Figure 70 Perovskite/silicon monolithic tandem solar cell evolution — **65**
Figure 71 Effect of spiro-OMeTAD thickness on top and bottom cell performances — **66**
Figure 72 Effect of the thickness of IZO on the performances of the top cell (on the left) and those on the bottom one (on the right), with a 200 nm thickness of the spiro-OMeTAD — **67**

List of tables

Table 1 The highest efficiencies of the perovskite-based solar cells —— **8**
Table 2 Characteristics of the perovskite/perovskite tandem solar cells developed in monolithic and in four-terminal configuration, as well as those of the constituent sub-cells, with an active area of 0.2 cm^2 —— **35**
Table 3 Characteristics of tandem solar cells with active area of 1 cm^2 —— **35**
Table 4 Comparison between the perovskite solar cell with the gold and the transparent-contact —— **39**
Table 5 The characteristics of the stand-alone CIGS solar cell when it is used in tandem configuration —— **40**
Table 6 The performances and efficiencies of the sub-cells, as well as the efficiency of the four-terminal tandem solar cell —— **42**
Table 7 The performances of the reference cells —— **51**
Table 8 The performances of the tandem solar cell —— **52**
Table 9 Characteristics of the perovskite solar cells in both illumination directions and in both scan directions —— **54**
Table 10 Characteristics of the silicon heterojunction bottom cell —— **56**
Table 11 The characteristics of the tandem solar cell for both aperture area, without antireflection layer —— **57**
Table 12 The characteristics of the tandem solar cell for both aperture areas, with the antireflective —— **58**

https://doi.org/10.1515/9783110760613-206

General introduction

The demand for electricity is growing steadily, and conventional energies are polluting and exhaustible, hence the interest in developing renewable energies. Indeed, several photovoltaic technologies have emerged. They are classified into four major generations [1].

The first generation, which includes monocrystalline and polycrystalline silicon technologies, represents 90% of the global market share [2]. Thin-film solar cells constitute the second-generation technologies and have shown interesting performance with lower costs than their predecessors. The third generation brings together the technologies of organic solar cells, such as dye cells, polymer cells, and perovskite solar cells. The last generation is called hybrid and contains multijunctions, among others.

The tandem configuration is an attractive alternative, which was proposed to further improve the performance of these solar cells at reasonable costs, while enhancing their competitiveness with conventional power sources [3].

The recently emerging perovskite solar cells show great potential and are considered to be the future of photovoltaics. Today, their record efficiency exceeds 25%, thanks to the wide choice of materials and to the simple, low-temperature process.

Since 2014, perovskite solar cells have demonstrated that they are perfectly suited to sub-cells over solar cells of other technologies, thanks to their compatible characteristics. In addition, these tandem solar cells based on perovskite sub-cells have shown a significant improvement in efficiency. Today, perovskite–silicon solar cells have achieved a record efficiency of 29.1%, and research continues to further records and innovations.

The aim of this book is to present the state of the art of perovskite solar cell technologies in single junction or in tandem configuration. The first chapter is devoted to perovskite solar cells, their characteristics, their structures, and principles of fabrication, as well as their scientific progress. The second one is dedicated to perovskite solar cell fabrication methods, including solvent process deposition, vapor deposition, and combined processes. The third chapter is about the perovskite/perovskite tandem solar cells and the perovskite/copper indium gallium di-selenide tandem solar cells, while detailing the fundamental of the tandem device and the best tandem configurations. Finally, the fourth chapter deals with the most attractive combination that interests several researchers and industrials, which is the perovskite/silicon heterojunction solar cells. This class of solar cells will be detailed from the principle to the most efficient devices.

https://doi.org/10.1515/9783110760613-001

Chapter 1
Perovskite-based solar cells

1 Introduction

Several materials are under development in order to further make photovoltaic technologies competitive with conventional energies. This chapter is dedicated to perovskite-based solar cells, with a focus on their component materials, characteristics, structures, and principle of operation, as well as their scientific progress.

2 Perovskites

Since 2009, hybrid perovskites, consisting of both organic and inorganic groups, have become attractive absorbers for third-generation solar cells with high efficiencies exceeding 25% [4, 5].

Perovskites refer to a family of materials of formula ABX_3 with a crystal structure similar to that of calcium titanium oxide ($CaTiO_3$) called perovskite. Hybrid perovskites are a subfamily of the ABX_3 perovskites. They are the result of the combination of the organic cations A of type ($R\text{-}NH_3$), a divalent metal B of group IV smaller than A, and a halide X or combination of halides, forming a three-dimensional (3D) inorganic structure that contains the organic cation [6].

3 Crystal structure of the hybrid perovskite

As shown in Figure 1, in the perovskite structure, B is a metal cation forming BX_6 octahedra with X anions. These octahedra touch at the vertices forming a 3D network. Cations A are located in the interstices between the BX_6 octahedra [7]. The crystalline structure of the perovskite depends strongly on the elements used [7, 8].

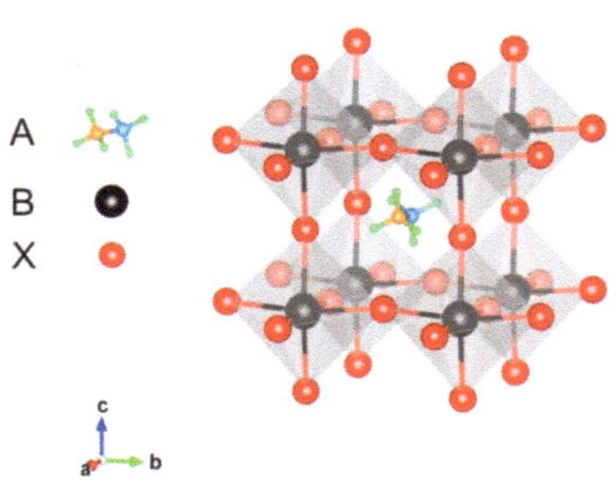

Figure 1: Crystalline structure of the hybrid perovskite. Source: Reprinted with permission from Dian Wang, Matthew Wright, Naveen Kumar Elumalai, Ashraf Uddin, Stability of perovskite solar cells, Solar Energy Materials and Solar Cells, Vol. 147, p. 255–275, Copyright (2016), with permission from Elsevier.

https://doi.org/10.1515/9783110760613-002

Indeed, the atoms and molecules used to form the hybrid perovskites are those who obey to the Goldschmidt tolerance factor, according to the equation:

$$t = \frac{(R_A + R_X)}{\sqrt{2}(R_B + R_X)}$$

where R_A, R_B, and R_X are the ionic radii of A, B, and X, respectively.

Indeed, the cubic structure is the most stable perovskite structure with a tolerance factor between 0.9 and 1, which is well suited for solar applications. Furthermore, for a tolerance between 0.8 and 0.9, the perovskite structure becomes distorted, giving less symmetric structures. However, when the tolerance factor becomes less than 0.8 or higher than 1, a nonperovskite structure is formed [9, 10].

Indeed, the atoms and compounds that meet the requirements of their crystalline structure and offer optical bandgaps suitable for semiconductors are listed below [4, 6, 9]:

A: Methylammonium $CH_3NH_3^+$ (MA), formamidinium $HC(NH_2)_2^+$ (FA), and cesium (Cs)
B: Lead (Pb) and tin (Sn)
X: I, Br, Cl, or mixture of two halides

4 Characteristics of perovskites

The properties of perovskite depend on the materials used and their fabrication history. On the other hand, the family of hybrid perovskites has some striking characteristics, namely:

- Absorption of wavelengths up to 800 nm as shown in Figure 2 [7, 11, 12]
- A direct bandgap between 1.30 and 2.23 eV [4, 7, 11, 13]
- Possibility of adjusting their optical bandgaps by an appropriate selection of mixed cations and halides as shown in Figure 3 [7, 12]
- Good separation of electrical charges, which explains their high open-circuit voltages (V_{oc}) up to 1.31 V [14]. Indeed, thanks to the spontaneous electric polarization of the perovskites, internal electric fields are induced, which makes it possible to separate the photoexcited charge carriers, while reducing their recombination [7].
- High charge-carrier mobilities, of the order of 7.5 cm^2/Vs for the holes and 12.5 cm^2/Vs for electrons, measured at terahertz TH_Z frequencies [7]
- Good light absorption, which explains the high short-circuit current density (J_{sc}) values, up to 23 mA/cm^2 [7]
- Large absorption coefficient, which is of the order of 1.5×10^4 cm^{-1} at a wavelength of 550 nm [6]
- High diffusion lengths for holes and electrons, reaching 100 nm for methylammonium lead iodide ($MAPbI_3$) and exceeding 1 μm for $MAPbI_{3-x}Cl_x$ [6, 7]

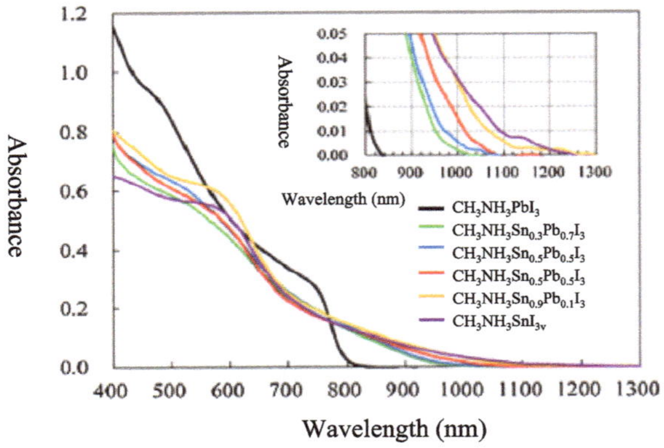

Figure 2: The absorption spectrum of the perovskite $CH_3NH_3Pb_{(1-x)}Sn_xI_3$.
Source: Reprinted with permission from Reprinted with permission from Q. Chen, N. De Marco, Y. Michael, T. Song, C. Chen, and H. Zhao, "Under the spotlight: The organic – inorganic hybrid halide perovskite for optoelectronic applications," Nano Today, vol. 10, no. 3, p. 355–396, 2015.

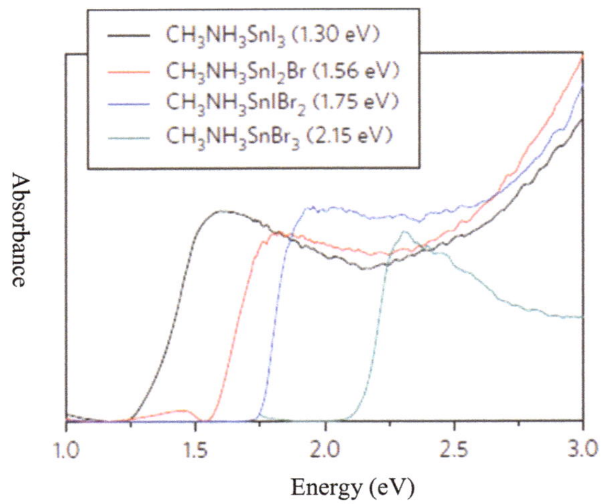

Figure 3: Optical bandgap for the perovskite $CH_3NH_3SnI_{3-x}Br_x$.
Source: Reprinted with permission from Springer Nature Customer Service Centre GmbH: Nature, Nature Photonics, Lead-free solid-state organic–inorganic halide perovskite solar cells, HAO, Feng, STOUMPOS, Constantinos C., CAO, Duyen Hanh, et al., 2014, vol. 8, no 6, p. 489–494.15. Copyright © 2014, Nature Publishing Group.

5 Structure of solar cells based on perovskites

5.1 Evolution of the structure of perovskite-based solar cells

Perovskite materials were primarily used in dye-sensitized solar cells (DSSCs) to replace conventional dyes, leading to perovskite-sensitized solar cell [15]. They showed a higher performance compared to the ruthenium-N719 dye used in DSSCs [16]. However, the use of liquid electrolytes in this type of solar cells, unfortunately, affects their stability. Moreover, the corrosive nature of the redox couple can also cause damage to the devices [17]. Furthermore, leakage and solvent evaporation can occur, which hinders the progress of this technology. Later, the meso-superstructured solar cell (MSSC) was proposed to reduce losses in the DSSCs, in which a very thin perovskite layer envelops the surface of mesoporous TiO_2 or Al_2O_3 films [18]. In this structure, an open-circuit voltage of 1.1 eV was achieved for a perovskite absorber with a band gap of 1.55 eV. Nevertheless, shunt and series resistances of MSSC were an obstacle to the performance of the device and remain to be optimized [18]. Afterward, the mesoporous structure was developed, where the perovskite covers the surface of the TiO_2 nanoparticles (NPs) while filling the pores of the mesoporous TiO_2 film [19]. In this structure, the collection of electrons became easier, since the contact area between the electron transporting layer (ETL) and the perovskite increased [20]. Moreover, it exhibits less hysteresis and better performance than planar architecture [21–23]. Nevertheless, the fabrication of the mesoporous oxide and the compact TiO_2 requires high temperature (T) up to 500 °C [24], which could raise the manufacturing cost of this kind of solar cells.

Finally, the planar perovskite solar cells were developed, where the perovskite layer is sandwiched between an n-type and p-type layers [4]. This structure does not require mesoporous TiO_2 or high processing temperatures like the mesoporous structure [25, 26]. Nevertheless, the efficiencies of this kind of solar cells are similar to those of the mesoporous structure [24–26].

The evolution of the structure of perovskite-based solar cells is presented in Figure 4.

5.2 Description of the structure of the perovskite solar cell

The planar perovskite solar cell is similar to the p-i-n structure, as shown in Figure 5. It is composed of three main layers, namely:

a. A light absorber: hybrid perovskite [6, 7]
b. An electron transport layer: n-type semiconductor [6, 7]
c. A hole transport layer: p-type semiconductor [6, 7]

Besides the transparent conductive oxide (TCO)-coated glass and the metallic contact, other layers can be added to the standard structure to enhance the performance of the

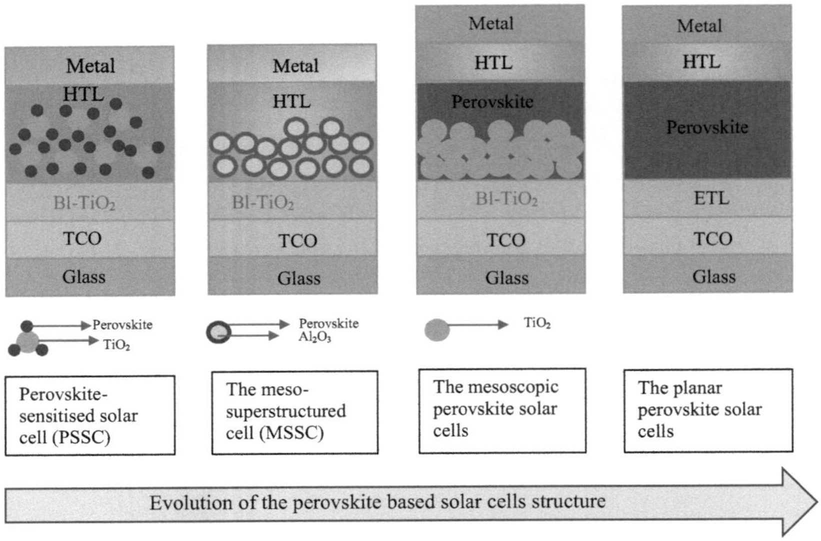

Figure 4: Evolution of the perovskite-based solar cell structure.

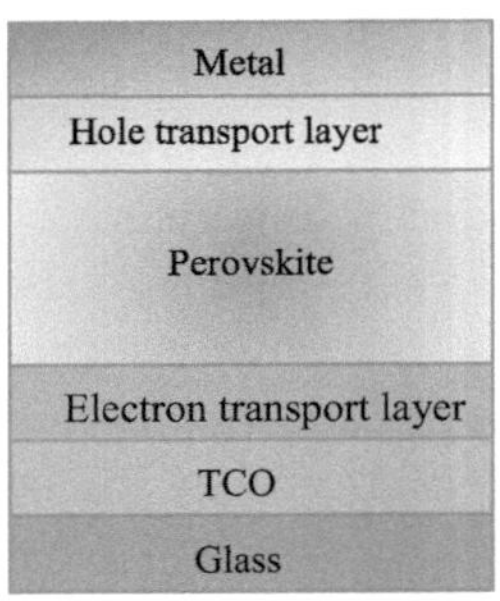

Figure 5: Standard structure of perovskite solar cell.

perovskite solar cells. Indeed, antireflective coatings were recommended to reduce losses, namely, magnesium fluoride [27] and lithium fluoride [28]. These investigations demonstrated an improvement of the efficiency of the solar cell up to 3% [29].

5.3 The components of the perovskite solar cell

5.3.1 Absorbers

Several types of absorbers from the perovskite family have been used in solar cells. The combination between different cations (A), metals (B), and halides (X), constituting perovskite, gives various parameters and performances.

Indeed, research has shown that perovskites based on tin (Sn) are very unstable, due to the instability of Sn^{2+} ions that easily transform into Sn^{4+} [30]. In addition, these methylammonium tin halide ($MASnX_3$) perovskites have optical bandgaps between 1.2 and 1.4 eV [7] and achieved recently 10.18% as the highest reported efficiency for pure tin halide-based Perovskite solar cells (PSCs) [31], which is lower than those reported for perovskites based on lead. Indeed, methylammonium lead halides ($MAPbX_3$) have an optical gap between 1.5 and 1.8 eV and lead to an efficiency of up to 21.38% [7, 32]. This is why lead-based perovskites are best suited for solar cells [6].

In the beginning of the perovskite solar cell technology, only the single halides were investigated. Indeed, $MAPbI_3$ was the most studied perovskite absorber with a maximum efficiency of 10.9% at the end of 2012 [6]. Furthermore, the perovskite based on bromine (Br) is known by its high bandgap and its improvement of solar cell efficiency and stability [7]. However, the combination of two halides provided better performances [33]. Indeed, perovskites containing iodide and chloride demonstrated an efficiency of 12.7% with a high diffusion length and improved stability compared to those containing only iodide [34]. Recently, a reported efficiency of 19.30% was achieved [35]. Furthermore, to improve the standard perovskite composition $MAPbI_3$, a partial substitution of the iodine by the Br was investigated and showed an enhancement of the efficiency and stability since its first demonstration by Noh et al., where 12.3% was reached in 2013 [36]. Later, this mixed-halide perovskites combining iodine (I) and Br with the formulation of methylammonium lead bromide–iodide ($MAPbI_xBr_{3-x}$) having bandgaps between 1.48 and 2.23 eV [7] achieved high certified efficiency up to 16.2% [13].

The hybrid perovskite $MAPbI_3$ has an optical bandgap between 1.51 and 1.55 eV [7]. When the MA cation is replaced by the FA, the optical bandgap decreases between 1.43 and 1.48 eV [7] and leads to efficiencies higher than 21% [37]. However, the perovskite containing Cs as cation and Br as halide, with the formula $CsPbBr_3$ and a wide bandgap of 2.25 eV, resulted in low efficiency of 5.95% in 2015 and evolved to less than 10% [38]. This is why this combination of atoms is not ideal for photovoltaic cells. On the other hand, iodide $CsPbI_3$ has a bandgap of 1.73 eV and good properties suitable for photovoltaic cells [33]. Recently, the $CsPbI_3$-based perovskite solar cells reached an efficiency of 18.4%, which is still quite limited due to the high bandgap [39].

The combination of two different cations offers interesting performances. Indeed, the best perovskite solar cells contain mixed organic cations (MA and FA) and mixed halides (Br and I) [33]. The first mixed perovskite cation with the formula $(MA)_x(FA)_{1-x}PbI_3$ resulted in an efficiency of 14.9% [7], and recently, $FA_{0.92}MA_{0.08}PbI_3$ achieved better efficiencies up to 23.7% [40]. Furthermore, perovskite with mixed cation and halide offered an efficiency of 20.8% for $MA_{0.17}FA_{0.83}Pb(I_{0.83}Br_{0.17})_3$ [33] and up to 23.95% for $(MA)_x(FA)_{1-x}PbI_xBr_{3-x}$ [41].

In addition, $MAPbI_3$ perovskite doped with Cs reached an efficiency of 7.68% in the beginning [7], and in the end of 2016, an efficiency of 18.1% was achieved with

$Cs_{0.09}MA_{0.91}PbI_3$ [42]. However, when Cs is mixed with FA cation, an efficiency of 16.5% was reached [33], and recently, a record efficiency of 21.60% was reported [41].

One of the interesting approaches is to combine Cs with the two organic cations MA and FA to have a triple cation configuration (Cs/MA/FA), this approach yielded efficiencies of up to 21.1%, with a very particular combination; $Cs_x(MA_{0.17}FA_{0.83})_{(1-x)}Pb(I_{0.83}Br_{0.17})_3$ [33, 41], and 21.4% with $Cs_{0.05}(MA_{0.17}FA_{0.83})_{0.95}Pb(I_{0.84}Br_{0.16})_3$ as an absorber [40]. In addition, other combination of tree cations yielded the highest efficiency of 21.9% using $Cs_{0.05}FA_{0.81}MA_{0.14}PbI_{2.55}Br_{0.45}$ as the active layer [43]. Furthermore, this triple cation configuration provides thermally stable and reproducible perovskite films.

Recently, rubidium cation (Rb) received great attention since Rb-mixed perovskite offers enhanced efficiency and stability of the solar devices. Indeed, Parck et al. reported $Rb_x(FA)_{1-x}PbI_3$ perovskite solar cells with a PCE of 16.2% [44]. Furthermore, the incorporation of the Rb with Cs, MA, and FA cations leads to stabilized efficiency of 21.6% [45].

However, when MA is substituted by a larger cation such as ethylammonium (EA), the structure of the perovskite changes completely, from a 3D structure to a two-dimensional (2D) one, which is more resistant to exposure to solar radiation [7]. Table 1 summarizes the highest efficiencies of the perovskite-based solar cells mentioned in this section.

Table 1: The highest efficiencies of the perovskite-based solar cells.

Perovskite absorber	Efficiency	Reference
$MASnX_3$	10.18%	[31]
$MAPbX_3$	21.38%	[32]
$MAPb(Cl,I)_3$	19.30%	[35]
$MAPbI_xBr_{3-x}$	16.2%	[13]
$FAPbI_3$	21.07%	[37]
$CsPbBr_3$	9.65%	[38]
$CsPbI_3$	18.4%	[39]
$FA_{0.92}MA_{0.08}PbI_3$	23.7%	[40]
$MA_{0.17}FA_{0.83}Pb(I_{0.83}Br_{0.17})_3$	20.8%	[19]
$(MA)_x(FA)_{1-x}PbI_xBr_{3-x}$	23.95%	[41]
$Cs_{0.09}MA_{0.91}PbI_3$	18.1%	[42]
$FA_{0.95}Cs_{0.05}PbI_3$	21.6%	[41]
$Cs_x(MA_{0.17}FA_{0.83})_{(1-x)}Pb(I_{0.83}Br_{0.17})_3$	21.1%	[19, 41]

Table 1 (continued)

Perovskite absorber	Efficiency	Reference
$Cs_{0.05}(MA_{0.17}FA_{0.83})_{0.95}Pb(I_{0.84}Br_{0.16})_3$	21.4%	[40]
$Cs_{0.05}FA_{0.81}MA_{0.14}PbI_{2.55}Br_{0.45}$	21.9%	[43]
$Rb_x(FA)_{1-x}PbI_3$	16.2%	[44]
$RbCsFAMAPb(I,Br)_3$	21.6%	[45]

5.3.2 Electron transporting layer

The ETL is used, as its name indicates to transport electrons efficiently from the perovskite absorber layer to the corresponding electrodes. This layer is considered at the same time as a hole blocking layer from perovskite to the Fluorine-doped Tin Oxide (FTO) contact in the planar perovskite solar cell, as shown in Figure 6 [23].

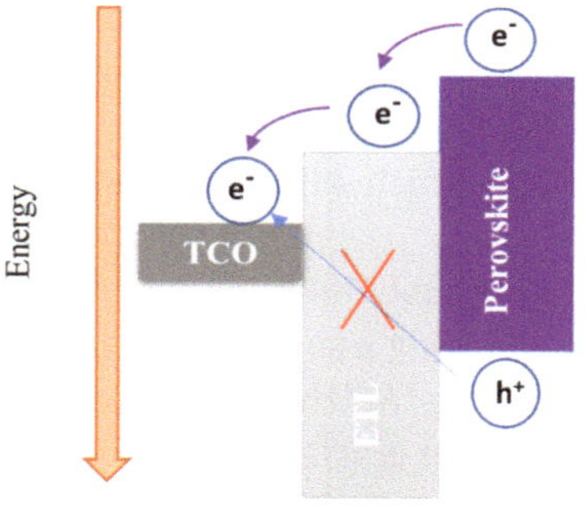

Figure 6: The role of the ETL on the movement of the electrons and holes.

For high-performance solar cells, materials used as ETL should meet a number of criteria, namely good electron mobility to facilitate the electron transportation, a high transparency in order to maximize light absorption in the perovskite materials, and compatible conduction band minimum (CBM) and valence band maximum (VBM) with those of the perovskite material in order to facilitate electron transfer and block hole transport from the perovskite layer as shown in Figure 6. Furthermore, the ETL should also limit reflection and light scattering, in order to increase the light harvesting of the perovskite absorbers [23].

Several materials have been used as ETL for the PSCs. They can be distinguished by their nature, notably metal oxide ETL such as titanium dioxide (TiO_2), zinc oxide (ZnO), tin dioxide (SnO_2), SiO_2, ZrO_2, and organic ETL like PEHT, PBCM [46].

5.3.2.1 Metal oxide ETL

The metal oxides were widely investigated as ETL in PSCs. TiO_2 is the most used one since the appearance of the PSCs. This metal oxide has a high bandgap of about 3.2 eV, which minimizes the parasitic absorption, and a good electron mobility of 1 cm^2/V s [23]. Furthermore, the TiO_2 CBM and VBM are lower compared to the energy levels of the perovskite absorber [23]. This promotes the transport of electrons from the perovskite to the ETL and at the same time works as a hole blocker as already explained.

Perovskite solar cells using TiO_2 as ETL have experienced a steady increase in efficiency starting with 5% [47] and reaching above 23% in the last years [40]. This efficiency improvement is related strongly to the fabrication method and the thickness of the TiO_2 and its structure, whether it is compact, mesoporous, or combination of both of them in the same configuration.

ZnO has attracted particular attention as an alternative to TiO_2 since it offers higher electron mobility up to 200 cm^2/V s [23]. Furthermore, it has the same bandgap as TiO_2 with a CBM and VBM lower than those of the perovskite absorber also [23]. In addition to these properties, ZnO is simple to synthesize at low temperatures which is suitable for flexible devices [48] and thus contributes to lower the costs. ZnO-based PSC yielded an efficiency of 15%; nevertheless, the thermal stability of the perovskite deposited on this oxide remains to be improved [49, 50]. Recently, the best efficiency of PSC with ZnO as ETL is reported to be around 17.7%, which still lower than that obtained with TiO_2 [23].

Moreover, Al-doped ZnO called AZO has an electron mobility higher than ZnO [48]. It was also used as an ETL, as well as TCO, thus allowing a simplified structure of solar cells as shown in Figure 7 for the case of glass/AZO/perovskite/spiro-OMeTAD/Au solar cell [49]. This type of structure offered an efficiency of 12.6% and good thermal stability compared to intrinsic ZnO. Furthermore, PSCs using AZO as ETL attained efficiencies up to 15.7% as reported by Liu et al. [23], and 17.6% with the following configuration indium tin oxide (ITO)/AZO/$MAPbI_3$/spiro-OMeTAD [48].

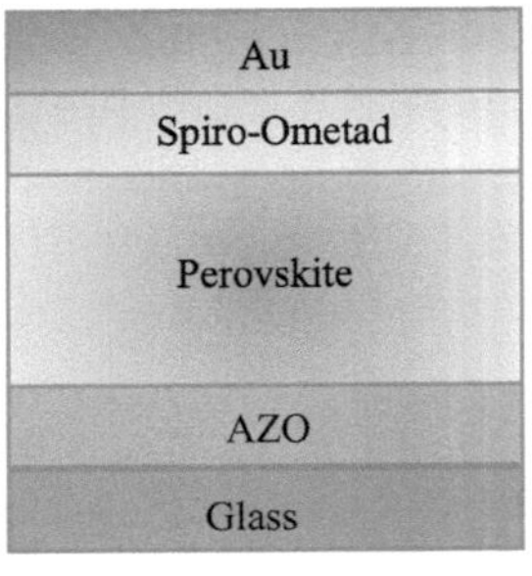

Figure 7: Perovskite solar cell structure containing AZO as TCO and electron transporting layer.

Besides, SnO_2 has similar or improved performance compared to TiO_2 [49]. Nevertheless, solar cells containing titanium oxide as the electron transporting material (ETM) degrade by ultraviolet radiation. However, SnO_2 makes it possible to have stable solar cells during exposure to solar radiation [51], with high efficiency up to 23.2% [52].

5.3.2.2 Organic ETL

Besides metal oxides, organic materials have also been investigated and used as ETL, namely, fullerene and its derivatives. Indeed, C70 and C60 ETL-based perovskite solar cells achieve a reported efficiency of 10.4% and 11.4%, respectively [48]. Other researchers demonstrated higher efficiency of 13.6% and 15.1%, respectively, for C70- and C60-based perovskite solar cell [23, 48]. Moreover, the other derivative of the fullerene called [6,6]-phenyl C61 butyric acid methyl ester (PCBM) is considerably used as ETL for PSCs with good efficiency up to 20.20% [41]. The PCBM is commonly used with PEDOT:PSS as HTL in the PSC structure [48].

Nevertheless, the electron mobility in C60 and PCBM is lower than TiO_2. Furthermore, the positions of the valence band of both of them are higher than that of TiO_2 leading to lower hole blocking ability [48]. In addition, the fullerene molecules of PCBM are sensitive to water and oxygen, which lead to degradation of the device [48]. However, it can be used as a passivation layer for inorganic-based ETL: TiO_2, ZnO, and SnO_2. The use of this passivation layer demonstrated higher efficiency and lower hysteresis effect [23].

5.3.3 Hole transporting layer

A hole transporting layer (HTL) is a thin film deposited on the perovskite absorber in order to collect and transport holes from the perovskite layer to the back contact of the solar cells.

The use of hole transporting materials makes it possible to improve the performance of the perovskite solar cells. In fact, a solar cell without an HTL gave an efficiency of 7.3%, with a gold (Au) film as a rear contact [7]. However, when the (2,2′7,7′-tetrakis(*N*,*N*-di-*p*-methoxyphenylamine)-9,9′-spirobifluorene) (spiro-OMeTAD) was added to the structure of the perovskite solar cell, higher efficiencies on the order of 9.7% were reached [7].

To obtain high-performance solar cells, materials used as HTL should demonstrate a number of properties, namely high hole mobility, good light absorption in the visible and near-infrared spectrum, and adequate photochemical stability. In addition, the highest occupied molecular orbital energy levels should be slightly above the valence band energy of perovskite materials to facilitate the injection of holes at the interface, and the lowest unoccupied molecular orbital energy level should effectively block recombination between electrons in perovskite and holes in HTLs [53, 54].

Various organic and inorganic materials have been explored as HTLs toward efficient PSCs.

5.3.3.1 Organic HTLs

The spiro-OMeTAD has been the first material used as HTL for perovskite solar cells [40]. It is an amorphous organic p-type semiconductor with a large bandgap between 3 and 3.6 eV [55]. It is considered as the most efficient HTL with good efficiency and better stability, thanks to its hydrophobic nature [56].

Indeed, spiro-OMeTAD-based perovskite solar cells are in continuous progress, and they achieved high efficiency of 21.3% [53]. However, the high cost and synthesis difficulties are the major disadvantages related to this HTL material [53, 54]. Furthermore, to improve the efficiency of the spiro-OMeTAD-based perovskite solar cells, dopants are recommended to enhance the hole mobility and the conductivity of the HTL and thus the efficiency of the resulting solar cells [54]. Indeed, the molecules usually used to dope and boost the performance of the spiro-OMeTAD are lithium bis(trifluoromethanesulfonyl)imide salt (LiTFSI) and tris(2-(1H-pyrazol-1-yl)-4-tert-butylpyridine) cobalt(III)tri[bis(trifluoromethane) sulfonimide] (Co[t-BuPyPz]3[TFSI]3), also called FK 209 [54].

Poly[bis(4-phenyl)(2,4,6-trimethylphenyl)amine] (PTAA) was the first polymer used as HTL for perovskite solar cells, since it has high hole mobility of 10^{-2}–10^{-3} $cm^2/V\ s$ compared to other polymers, leading to an efficiency of 22.7% and improved stability compared to the PEDOT:PSS [40, 54]. However, it is very expensive (2,000 $/g) compared to the spiro-OMeTAD itself [54].

Another well-known conjugated polymer is the poly(3,4-ethylenedioxythiophene) (DEPOT:PSS) doped with polystyrene sulfonate (PSS), which is widely used as an HTL for perovskite solar cells, especially for inverted planar architectures [56]. It resulted in efficiencies of up to 18.1%, since it has a good conductivity and a work function compatible with the valence band of the perovskite, and it can be processed at low temperature [40, 54]. However, its low stability in ambient condition limits its use due to its hygroscopic property [40, 54], and it leads to a lower open-circuit voltage V_{oc} compared to spiro-OMeTAD [56].

Poly(3-hexylthiophene-2,5-diyl):(P3HT) is a low-cost polymer used as a hole transporting material in perovskite solar cells [56]. In addition, when gold nanoparticles (Au NPs) are added to the matrix of P3HT, the efficiency of the perovskite solar cell increases from 7.81% to 10.71% [56]. This result is explained by the increase of the electrical conductivity and the mobility of the carriers, which become four times higher than those of the film containing P3HT alone. Recently, 13% efficiency is achieved for P3HT-based PSCs [54].

5.3.3.2 Inorganic HTLs

Copper-based materials are one of the used inorganic HTLs for perovskite solar cell. Indeed, copper iodide (CuI), which is one of this class of cheaper HTL, resulted in an interesting fill factor and a more stable current density than spiro-OMeTAD. However, the open-circuit voltage V_{oc} of the CuI-based PSCs is lower than that of spiro-OMeTAD based solar cells, which limits the efficiency of the solar cell to only 8.3% [57, 58].

In addition, cuprous oxide (Cu_2O) has a high hole mobility, and it is considered as the low-cost and ecological material [54]. Moreover, the use of Cu_2O and cupric oxide (CuO) as HTL in the PSCs showed higher efficiency of 13.35% and 12.6%, respectively [57]. Nevertheless, their relatively low bandgap (1.2 and 2.1 eV for CuO and Cu_2O, respectively) leads to optical losses for the solar cells [40]. Furthermore, CuO_2 has demonstrated improved stability for perovskite solar cells, as they retain 90% of their initial efficiency after 70 days of storage [57].

Moreover, nickel oxide (NiO) is considered as a low-cost material and abundant one, which exhibits wide bandgap, high chemical stability, and good transparency leading to high efficiency up to 16.4% for inverted planar PSCs [54, 59]. Furthermore, copper-doped NiO_x (Cu:NiO_x) exhibited high conductivity of 1.25×10^{-3} S/cm and resulted in efficiencies of 17.74%, which is higher than those based on undoped NiO [53].

Furthermore, copper thiocyanate (CuSCN) was one of the first inorganic HTLs used for perovskite solar cells. Indeed, Gratzel et al. reported a great enhancement of the efficiency of the PSCs using CuSCN as an HTL. They reported an efficiency of 12.4% in 2014, compared to 9% of the reference solar cell without HTL [54]. In addition, Nazeeruddin et al. also reported a significant improvement from 9.5% to 16% for the PSCs without HTL and with CuSCN, respectively [54]. Moreover, CuSCN offers high transparency in both visible and infrared regions in addition to a high mobility, resulting in efficiencies up to 20.4% [57]. In addition, perovskite solar cells based on CuSCN are highly stable compared to those based on spiro-OMeTAD as they maintain 60% of their initial efficiency compared to only 25% for spiro-OMeTAD-based perovskite solar cells in the same conditions [54].

In brief, inorganic HTL-based perovskite solar cells lead to high efficiency and improved stability; however, they require high-temperature processes compared to those based on the organic HTLs [53].

5.3.4 Metal contacts

Au and silver (Ag) have high conductivity and reflectivity. They are the most frequently used metal contacts as a back contact in perovskite solar cells. Nevertheless, Ag is reactive with perovskite absorbers [60].

5.3.5 Substrates

Perovskite solar cells have offered the possibility to use substrates other than glass, and more specifically transparent flexible substrates, such as poly(ethyleneterephthalate) (PET)/ITO, polyethylenenaphthalate (PEN)/ITO, PET/IZO, PET/AZO/Ag/AZO [7].

5.4 Structures of the perovskite solar cells

The stacking of solar cell components and their natures distinguish one solar cell structure from another. Indeed, there are structures without transport layer neither for electrons nor for holes [7]. Since perovskites are ambipolar, they allow the conduction of both electrons and holes [6]. In addition, perovskite solar cells can have two configurations such as p-i-n or n-i-p, as presented in Figure 8 [6, 7]. However, the most efficient perovskite solar cells are those with a p-i-n configuration [7].

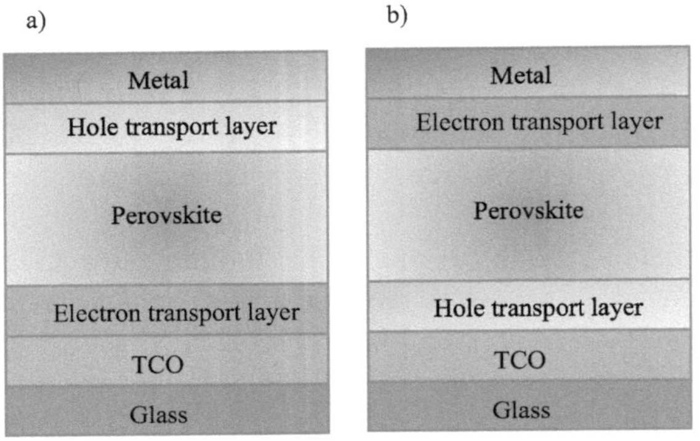

Figure 8: Perovskite solar cell structures (a) n-i-p and (b) p-i-n configurations, respectively.

Furthermore, perovskite solar cell structures can be classified according to the types of ETL and HTL used. According to this classification, there are four structures, namely:

5.4.1 Inorganic n-type/perovskite/organic p-type

This structure gave an efficiency of 14.1%, using TiO_2 as n-doped inorganic ETL and spiro-OMeTAD as p-type organic HTL [19]. An efficiency of 17.9% was obtained using ZnO as n-layer and polytriaryl amine (PTAA) as p-layer [5].

5.4.2 Organic n-type/perovskite/inorganic p-type

Another structure, where methyl[6,6]-phenyl-C61-butanoate (PCBM) was used as the n-layer and NiO as p-layer, reached an efficiency of 9.51% [61].

5.4.3 Organic n-type/perovskite/organic p-type

This configuration gave an efficiency of 16%, with PCBM as n-doped organic layer and PEDOT:PSS as doped organic layer [62].

5.4.4 Inorganic n-type/perovskite/inorganic p-type

In this configuration, TiO_2 is the n-doped inorganic layer and CuSCN is the p-doped layer. This solar cell reached an efficiency up to 12.4% [63].

6 Operating principle of a perovskite solar cell

During the illumination of the perovskite solar cells with photons, electrons are excited from the valence band to the conduction band of the absorber (active) layer leaving holes behind. Since the binding energy of excitons in perovskites is rather small, the excitons dissociate shortly, thus forming free charge carriers. The photogenerated excess electrons and holes diffuse and are selectively transferred through ETL and HTL to the bottom and top electrodes, as presented in Figure 9 [6].

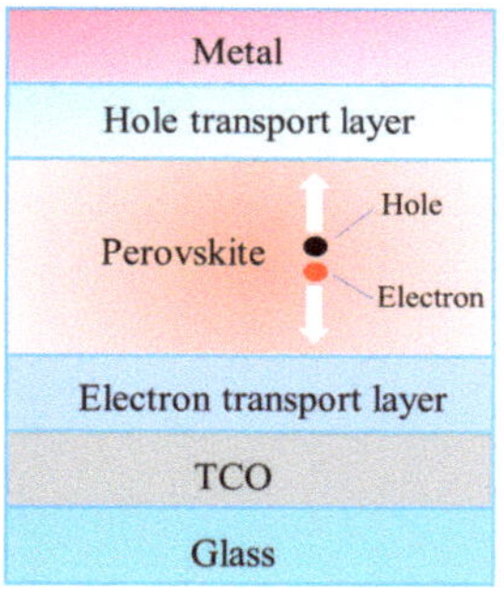

Figure 9: Operating principle of a perovskite solar cell.

7 The evolution of the perovskite solar cell efficiency

The perovskite materials were introduced into solar cells in 2009, and since then have been in continuous development depending mainly on the chemical composition of the layers and its fabrication method, in addition to the structure of the whole solar cell.

Indeed, in 2009, Miyasaka et al. found an efficiency of 3.8%, for the perovskite $MAPbI_3$ deposited on nanocrystals of TiO_2 [4]. In 2011, Park et al. reported higher efficiency of 6.5% [64]. At the end of 2012, Gratzel et al. added spiro-OMeTAD to $MAPbI_3$, which reached 9.7% [4]. Later, Snaith et al. replaced the n-type TiO_2 ETM with inert Al_2O_3 alumina (scaffold) and achieved an efficiency of 10.9% [7]. Then, mixed-halide perovskite was investigated in 2013 by Seok et al., and reported 12.3% efficiency for $CH_3NH_3PbI_{3-x}Br_x$ [64]. Subsequently, in 2014, Jeon et al. used PTAA as HTL for mixed-halide perovskite solar cells and found an efficiency of 16.2% [13]. In the end of the same year, an efficiency of 19.3% was reached [65]. Next year, an efficiency of 20.1% was reported by Yang et al. Furthermore, in 2016, Saliba et al. certified a new record efficiency of 21.1% [33]. With continuous efforts, a certified recording of 22.1% was reached [66]. Later, according to the NREL charts, a remarkable progress of PSCs was made achieving 23.3% and 23.7% in 2018 and 2019, respectively. Today, the current highest certified efficiency of PSCs has reached 25.2% [5].

Figure 10 shows the evolution of the perovskite-based solar cells already presented in this section.

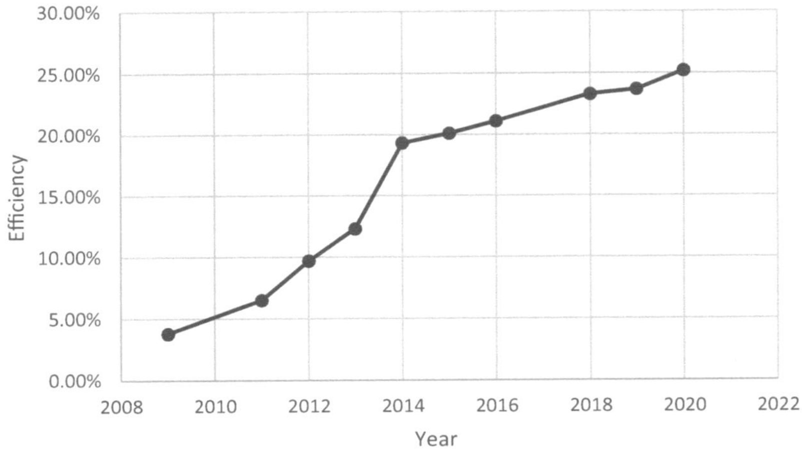

Figure 10: The evolution of the perovskite-based solar cells.

8 Conclusion

The hybrid perovskite absorbers demonstrate great characteristics and show advantages to further enrich and develop the field of renewable energies. Enormous progress of the performance of the perovskite solar cells has been achieved due to the great and persistent effort made by researchers from around the world. Thanks to them, the efficiency has now reached 25.2% and the perovskite solar cell becomes competitive to the silicon one.

Chapter 2
Perovskite solar cell fabrication methods

1 Introduction

Several fabrication methods have been used and studied for each layer of the perovskite solar cells. They require less means and lower temperatures which explain why their cost is lower than that of crystalline silicon ones [67]. In this chapter, several methods for preparing perovskite absorber will be presented.

2 One-step spin-coating method

The spin-coating technique is considered as the most frequently used method since it is fast, simple, and cheaper [60]. Indeed, to deposit perovskite thin film, a dozen microliters of the precursor's solution is dripped on a substrate fixed on the spin-coater and then annealed as schematically depicted in Figure 11. This technique requires few seconds and low-cost precursor material compared to other technologies. Perovskite absorber is generally deposited under an inert atmosphere or glove box, by spin coating in one step, from a molar precursor solution (1:1) of methyl ammonium iodide (MAI) with a formula of CH_3NH_3I and lead iodide (PbI_2), solubilized in a solvent such as γ-butyrolactone or dimethylformamide (DMF) or a mixture of dimethylsulfoxide (DMSO) and DMF [6].

This procedure is considered to be better in terms of step and time minimization, but it does not control either the growth kinetics of the crystals or the morphology of the coating, which explains why the resulting performances are less interesting than those produced by spin coating in two steps [4].

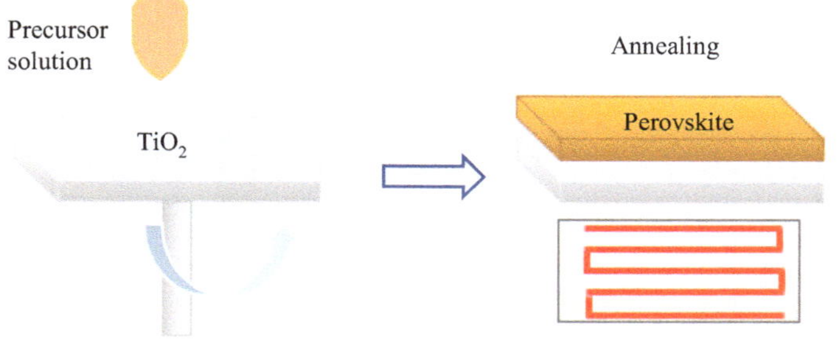

Figure 11: One-step spin-coating method.

https://doi.org/10.1515/9783110760613-003

3 The hot-casting technique

The hot-casting technique consists in using a precursor solution heated to a temperature of 70 °C, dripped on a glass substrate heated to a temperature close to or greater than that of the boiling point of the solvent, and then spin-coated as shown in Figure 12 [68].

This technique favors the formation of perovskite grains larger than those obtained by the other methods using the post-annealing, with a size in the range of 1–2 mm compared to 1–2 µm only. In addition, the perovskite films prepared by this technique cover the entire surface of the substrate without pinholes [68].

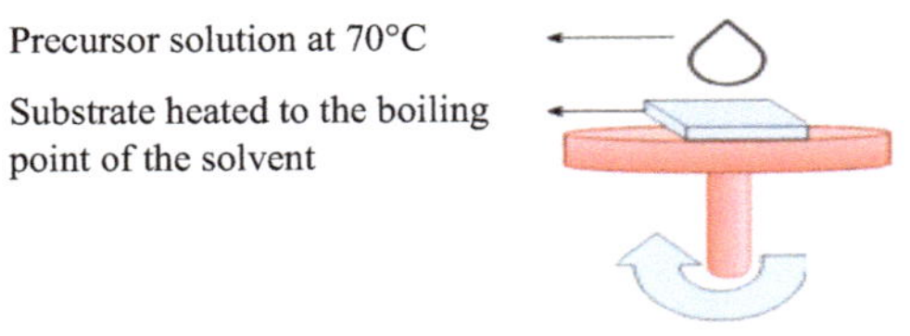

Figure 12: The hot-casting technique.

4 The anti-solvent method

The anti-solvent method or rapid crystallization technique has been proposed to produce a flat and uniform $MAPbI_3$ film with well-developed grains [4].

When the solvent *N*,*N*-dimethylformamide (DMF) is used in one-step deposition, the result is a nonuniform perovskite film containing holes resulting from the high boiling point of the solvent (153 °C), thus allowing a slow crystallization. Nevertheless, in this technique, a solution containing a mixture of DMF and $MAPbI_3$ is first applied by spin coating to the TiO_2 layer, and after a specific time, the anti-solvent (toluene) is rapidly added to the substrate, to promote rapid nucleation and growth of $MAPbI_3$ by reducing the solubility of $MAPbI_3$ in the mixed solvent. This method is illustrated in Figure 13.

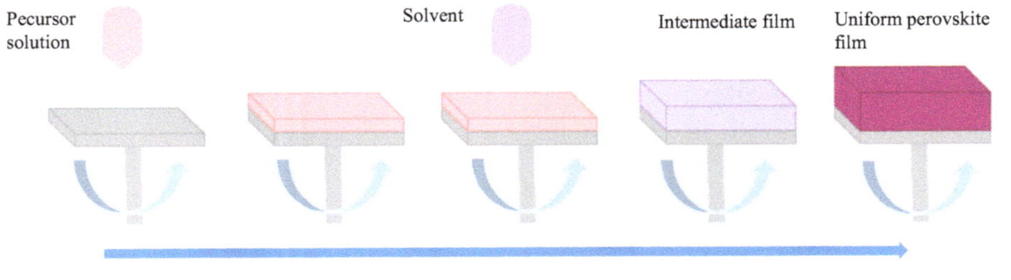

Figure 13: The anti-solvent method.

5 The two-step spin-coating technique

In this technique, as shown in Figure 14, a solution of PbI_2 is first applied by spin coating to the mesoporous film of TiO_2. The PbI_2-coated film is dried at 40 °C for 3 min and then at 100 °C for 5 min. Then a solution of MAI solubilized in 2-propanol (IPA) is deposited by spin coating on this dried PbI_2 film for 20 s. Finally, the film is heated at 100 °C for 5 min [12].

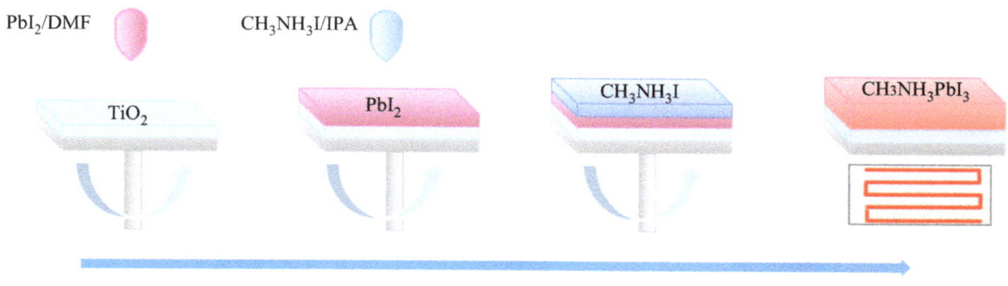

Figure 14: The two-step spin-coating technique.

The performance of perovskites deposited by the two-step process is higher than those deposited by the one-step process. This is due to the morphology of the resulting perovskites [12]. Indeed, this technique made it possible to have an efficiency of 15% [7], for the following structure: PEDOT:PSS/$CH_3NH_3PbI_3$/PCBM/C60/BCP/Al.

6 The dual-source vapor deposition technique

Vapor deposition technique has also been used for perovskite thin film deposition, either in one or two steps [4]. This technique offers flat, uniform films with better quality than perovskite deposited by spin coating, since there is no solvent in the process, which avoids the formation of pinholes resulting from the evaporation of the solvent [4, 6].

In dual-source vapor deposition technique, the two precursors containing lead and MA are in powder form inside the vacuum chamber and they are evaporated simultaneously to be deposited on the substrate, as shown in Figure 15 [4].

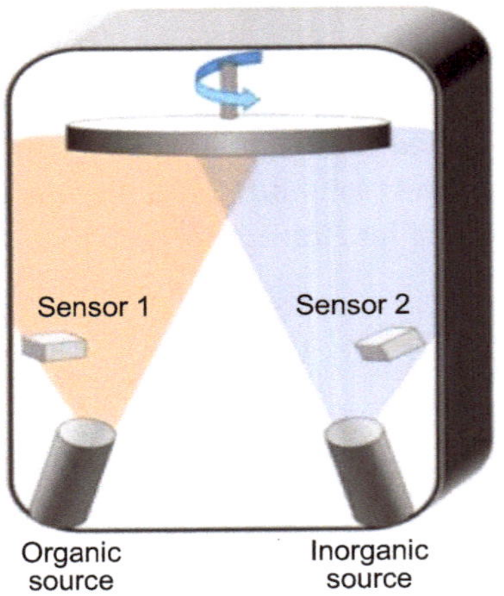

Figure 15: The dual-source vapor deposition technique.
Source: Reprinted with permission from N.-G. Park, "Methodologies for high efficiency perovskite solar cells," vol. 3, p. 15, 2016, doi: 10.1186/s40580-016-0074-x.

7 The sequential dual-source vapor deposition technique

In the sequential dual-source vapor deposition technique, the PbI_2 is deposited first by thermal evaporation followed by MAI on a preheated substrate between 65 and 85 °C, as schematized in Figure 16. The deposition is made at a speed of 0.03 nm/s, in a chamber reaching a pressure of 1.23×10^{-4} mbar, leading to perovskite solar cell with 15% efficiency [7].

The interest of the vapor phase deposition in two steps stems from controlling the deposition rate of MAI, which is not controllable in the co-deposition and to avoid the problems resulting from solvents such as the slow crystallization of the

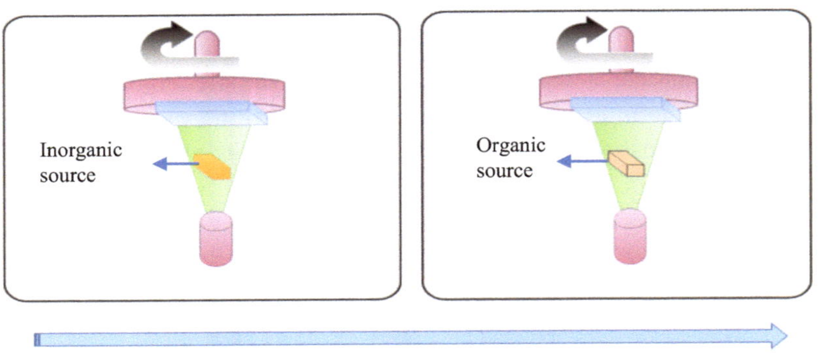

Figure 16: The sequential dual-source vapor deposition technique.

perovskite and the pinhole formation within the perovskite film [4, 6]. Furthermore, perovskite films deposited by the vapor deposition technique offer higher performance than those deposited by solution process; as an example, the $MAPbI_{3-x}Cl_x$ film deposited by vapor deposition technique exhibits higher charge mobility of 33 cm^2/V s compared to only 11.6 cm^2/V-1S-1 when deposited by solution process.

8 The adduct method

As one of the effective methods, the adduct approach has been proposed to prepare high-quality perovskite $MAPbI_3$ films. In effect, the interaction between DMSO solvent as Lewis base and PbI_2 as Lewis acid resulted in a clear adduct film which was converted to $MAPbI_3$ by removing DMSO with mild thermal treatments [4].

The equimolar mixture of PbI_2, MAI, and DMSO in DMF solvent is deposited on a substrate via the spin-coating method, then diethyl ether is added during the centrifugation, as schematized in Figure 17, leading to a transparent film which is directly indicative of the formation of the adduct product. The purpose of using diethyl ether is to remove only DMF to form the adduct film (1:1:1). The thermal removal of DMSO from the adduct film allows the control of the growth kinetics of $MAPbI_3$.

This method resulted in a high charge carrier mobility of up to 3.9×10^{-3} cm^2/V s (measured by photo-CELIV), unlike that measured for perovskite films deposited by the one-step spin-coating method, which does not exceed 3.2×10^{-4} cm^2/V s.

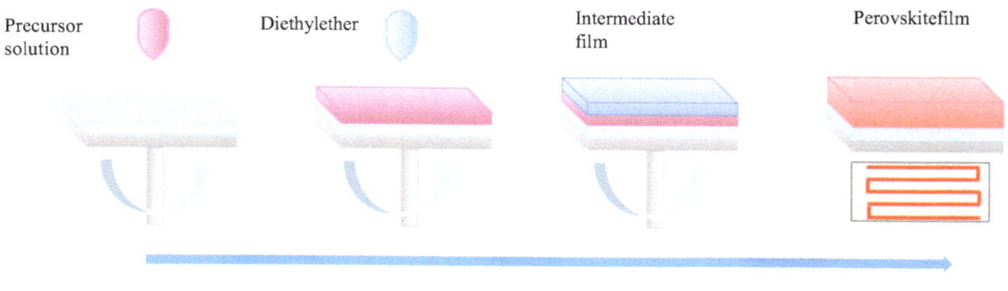

Figure 17: The adduct method.

9 The vapor-assisted solution process (VASP)

This method was developed to avoid the co-deposition of organic and inorganic species. It consists of treating the PbI_2 film, deposited on TCO coated with a compact TiO_2 layer, with the vapor-phase MAI. Then, the already deposited PbI_2 films are annealed in MAI vapor at 150 °C, in a nitrogen atmosphere for 2–4 h in order to form uniform perovskite films with large grains, as schematized in Figure 18 [4].

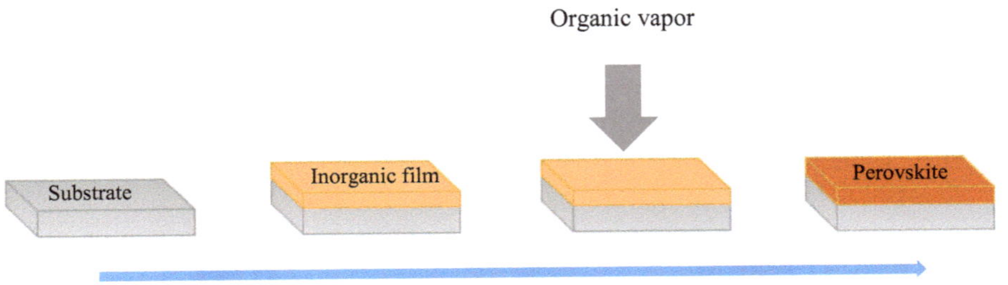

Figure 18: The vapor-assisted solution process (VASP).

Perovskite solar cells derived from this approach offer high efficiency of 20.1%, since the resulting perovskite films have a uniform grain structure with particle size up to micrometers with complete surface coverage and the 100% precursor transformation [6]. In brief, this technique is considered as a good process to get high-quality perovskite films and thus high-performance devices.

10 Ink-jet printing technique

Ink-jet printing is a noncontact technique to produce a specific pattern for many applications including PSCs with its different layers (perovskite, ETL, HTL, and contact) [70]. This technique consists of printing inks through the nozzles on the substrate, as schematized in Figure 19. These inks should be engineered to have a suitable solubility so as not to block the nozzles, which is considered as a recurrent issue in this technique.

The layers issued from this method have a homogenous thickness and controllable shape and size [70]. Furthermore, this new method guarantees high scalability,

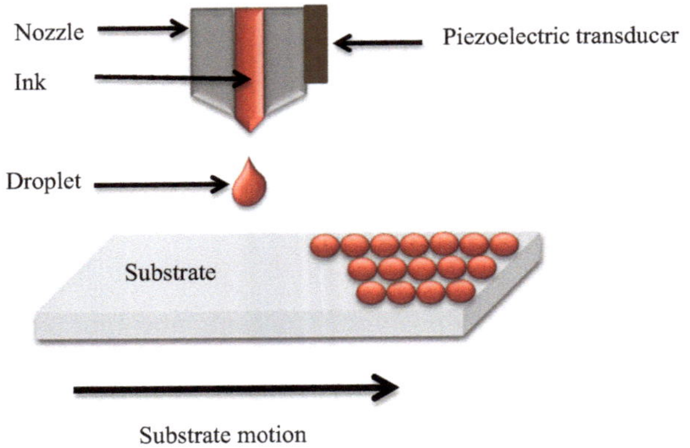

Figure 19: Ink-jet printing illustration.

less material wastage, fast deposition, which make it attractive for printing PSCs.

11 Techniques to improve the performance of perovskite

Today, improving the performance and stability of perovskite solar cells remains the main mission allotted to all researchers in the development of perovskite solar cells. In this section, several solutions have been proposed to enhance the efficiency and the stability of perovskite films and thus to improve the performance of the whole perovskite solar cells.

11.1 Techniques to improve perovskite efficiency

A good perovskite film guarantees a good coverage of the surface without pinholes to avoid short circuits between the different layers of the solar cells. Several techniques have been investigated to improve the perovskite thin film quality.

In this regard, a simple chemical approach has been developed by introducing an appropriate amount of acetic acid during the perovskite film preparation, leading to a homogeneous and continuous films, without holes, as shown in the images gathered in Figure 20 [71]. Moreover, the films are composed of grains with high crystallinity

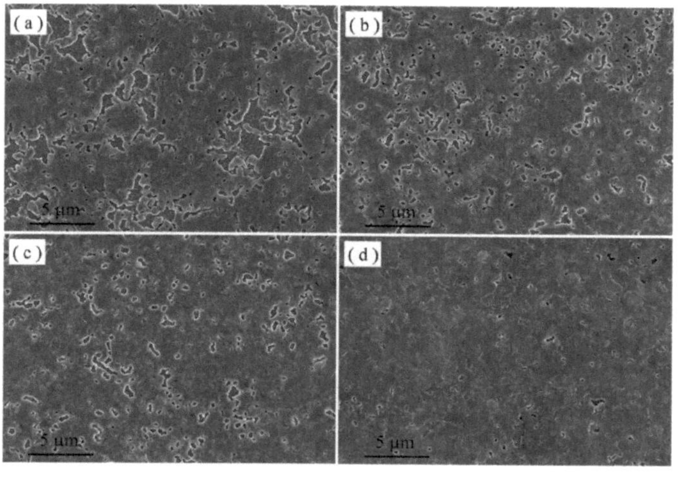

Figure 20: The images (a, b, c, d) correspond to the perovskite films deposited on FTO /compact-TiO_2 substrates from a precursor solution containing amounts of 0, 0.25, 0.5, and 1.0 M acetic acid, respectively. Source: Reprinted with permission from ZHAO, Qian, LI, G. R., SONG, Jian, et al. Improving the photovoltaic performance of perovskite solar cells with acetate. Scientific reports, 2016, vol. 6, no 1, p. 1–10.

and exhibit higher light absorption, more efficient electron transport, and faster electron extraction at the adjoining electron transport layer [71]. In addition, the perovskite solar cells based on these films have demonstrated a better efficiency up to 14.17% [71].

Furthermore, the quality of the perovskite thin film was improved by adding CH_3NH_3Cl to the precursor solution in one-step spin-coating technique [72]. Moreover, the use of 1% of 1,8-diiodooctane as an additive to the precursor solution demonstrated perovskite films with better performance [7]. In addition, the ratio between the two precursors PbI_2 and MAI affect also the quality of the deposited perovskite film. A 1:3 ratio resulted in a continuous film and a higher efficiency of up to 19% [7].

11.2 Techniques for improving the perovskite stability

Perovskite is considered as a fragile material, soluble in water, and therefore it degrades rapidly and irreversibly. This is one of the major obstacles to the commercialization of perovskite solar cells. Indeed, temperature, UV light, and moisture are considered as the major factors affecting the perovskite stability [60]. Several solutions have been proposed to protect and to make perovskite-based devices more stable.

Indeed, one of the solutions investigated is the auto-repair of photovoltaic cells in the darkness, which consists of putting the perovskite solar cells in the dark for 1 min, allowing the solar cell to recover their initial efficiencies, as demonstrated in Figure 21. Indeed, during the day, electrical charges are trapped in the defects of the crystal lattice, which disrupts the continuous flow of these carriers. The darkness then allows to evacuate these trapped charges, which has the consequence of recovering the initial efficiencies of the perovskite solar cells [73].

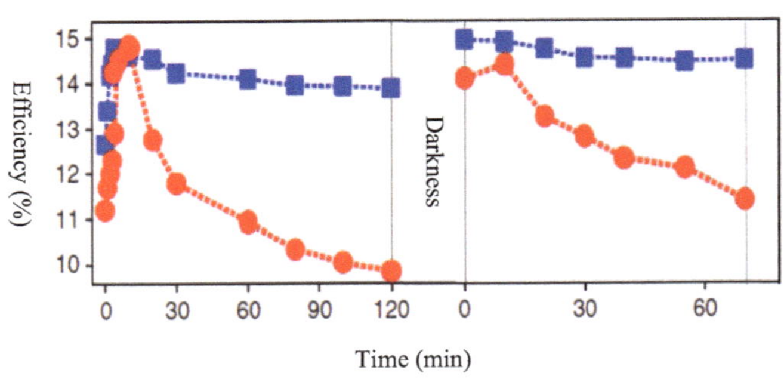

Figure 21: The blue curve represents the evolution of the perovskite solar cell efficiency as a function of time, tested in short circuit while the red one is tested in open circuit.
Source: Reprinted with permission from NIE, Wanyi, BLANCON, Jean-Christophe, NEUKIRCH, Amanda J., et al. Light-activated photocurrent degradation and self-healing in perovskite solar cells. Nature communications, 2016, vol. 7, no 1, p. 1–9.

Moreover, improving the crystalline quality of perovskite remains an interesting way to overcome the issue of the grain boundaries since they are potential pitfalls for electrical charges, which reduce the performance of perovskites over time. In this regard, developing large grains during the crystallization of perovskite at appropriate temperatures reduces the grain boundaries and the recombination [68]. Nie et al. demonstrated that substrates heated at 190 °C resulted in perovskite thin films with large grain size, as shown in Figure 22.

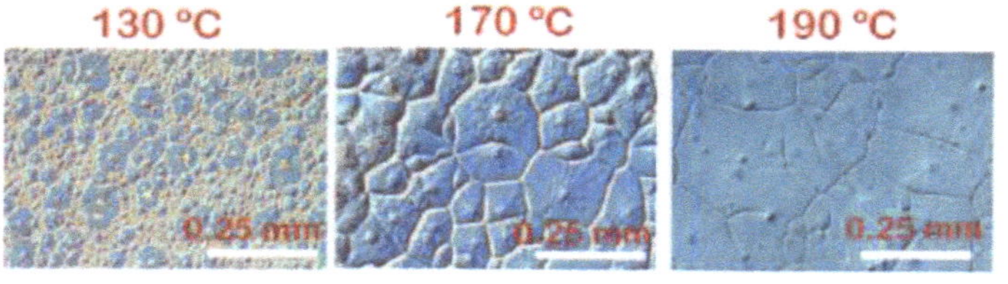

Figure 22: The size of grains as a function of the temperature of the substrate with the same precursor solution at 70 °C.
Source: Reprinted with permission from THE AMERICAN ASSOCIATION FOR THE ADVANCEMENT OF SCIENCE: Science, "High-efficiency solution-processed perovskite solar cells with millimeter-scale grains," NIE, Wanyi, TSAI, Hsinhan, ASADPOUR, Reza, et al., vol. 347, no. 6221, p. 522–525, 2015.

Moreover, the cesium (Cs) incorporation in the A position of the perovskite structure leads to a better stability and efficiency [30]. Indeed, Lee et al. substituted 10% of formamidinium (FA) by Cs leading to perovskite formula of $FA_{0.9}Cs_{0.1}PbI_3$ and thus the efficiency was improved by 10%, with better stability as shown in Figure 23.

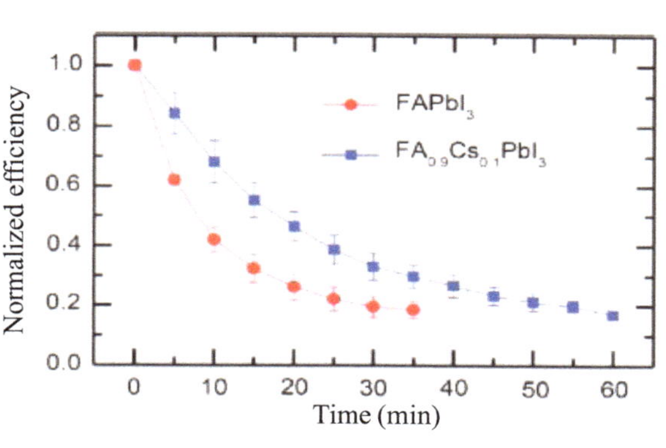

Figure 23: Effect of cesium on the stability of perovskite.
Source: Reprinted with permission from Dian Wang, Matthew Wright, Naveen Kumar Elumalai, Ashraf Uddin, Stability of perovskite solar cells, Solar Energy Materials and Solar Cells, Vol. 147, p. 255–275, Copyright (2016), with permission from Elsevier.

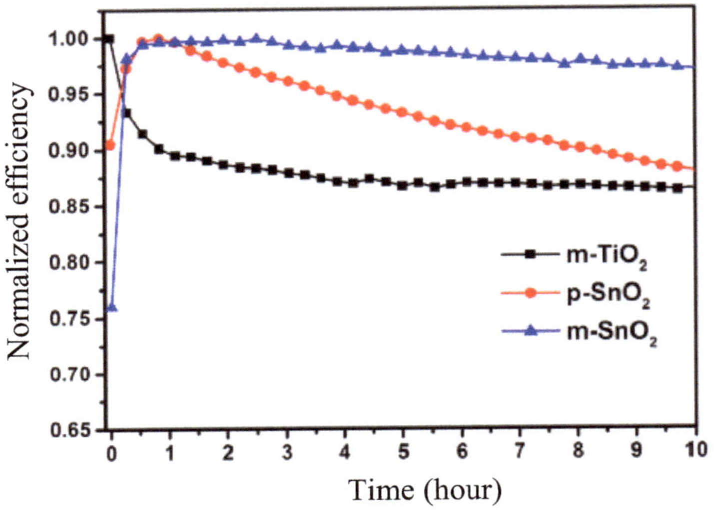

Figure 24: Effect of mesoporous TiO_2 and SnO_2, and planar SnO_2 on the stability of the perovskite solar cell.
Source: Reprinted with permission from ROOSE, Bart, BAENA, Juan-Pablo Correa, GÖDEL, Karl C., et al. Mesoporous SnO2 electron selective contact enables UV-stable perovskite solar cells. Nano Energy, 2016, vol. 30, p. 517–522.

Furthermore, using SnO_2 as the electron transporter for the perovskite solar cell instead of TiO_2 also improves the stability of the solar cell, as demonstrated in Figure 24. Indeed, perovskite solar cells with SnO_2 showed a stability of 700 h [30]. Whereas solar cells containing TiO_2 reached a stability of 500 h [19]. This may be explained by the intrinsic properties of each electron transporter. On the one hand, the exposure of TiO_2 to UV light induces a degradation of the perovskite solar cell efficiency [51]. On the other hand, SnO_2 has a wide bandgap and absorbs less UV light, making the perovskite solar cell more resistive to UV degradation than that based on TiO_2 [51]. In addition, SnO_2 has a higher electron mobility than TiO_2 and better electron injection [51], making it a potential competitor to the widely used TiO_2.

Moreover, to overcome the TiO_2 UV sensibility, other materials have been developed. Indeed, Zn_2SnO_4 was proposed as electron transporter layers instead of the TiO_2 and the resulting perovskite solar cell reached 13.3% efficiency and showed a 1-month stability without any encapsulation [74]. In addition, Sb_2S_3 has been used to reduce UV-induced degradation and the Sb_2S_3-based perovskite solar cell retained nearly 65% of its initial efficiency without encapsulation [75].

In addition, when the perovskite solar cell is encapsulated, it is automatically protected against degradation caused by the external factors. Figure 25 shows that encapsulated solar cells have greater stability than non-encapsulated solar cells.

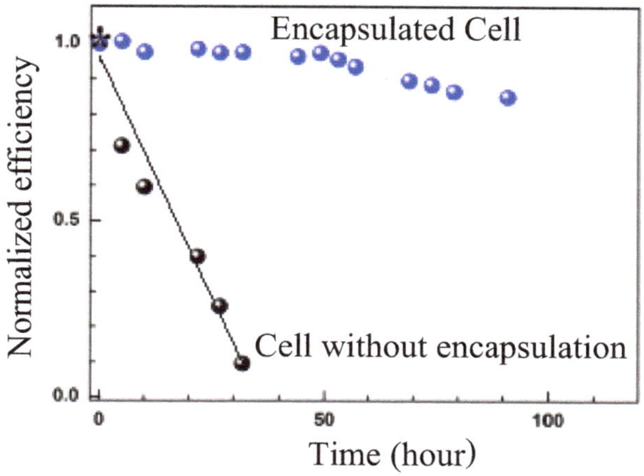

Figure 25: Effect of the encapsulation on the stability of the perovskite solar cell.
Source: Reprinted with permission from CUI, Jin, LI, Pengfei, CHEN, Zhifan, et al. Phosphor coated NiO-based planar inverted organometallic halide perovskite solar cells with enhanced efficiency and stability. Applied Physics Letters, 2016, vol. 109, no 17, p. 171103.

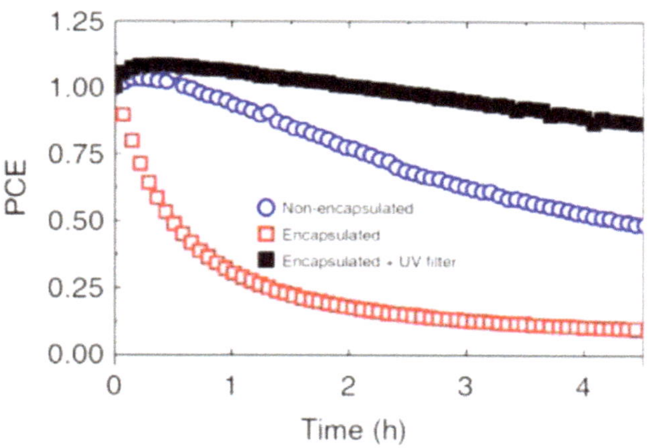

Figure 26: The effect of the encapsulation and UV filter on the perovskite solar cell degradation.
Source: Reprinted with permission from Stability of perovskite solar cells, Solar Energy Materials and Solar Cells, Vol. 147, Dian Wang, Matthew Wright, Naveen Kumar Elumalai, Ashraf Uddin, p. 55–275, Copyright (2016), with permission from Elsevier.

In fact, the materials used for encapsulation are hydrophobic and resistant to ultraviolet (UV) irradiation; thus, the penetration of moisture and the degradation by UV radiation are reduced [7, 8, 76].

Moreover, some researchers demonstrated higher protection from the degradation of PSCs by using both encapsulation and a UV filter. Figure 26 shows that the perovskite solar cells encapsulated without UV filter degraded more quickly than those without any encapsulation; however, those encapsulated with the UV filter present higher stability [77].

12 Conclusion

Perovskite solar cells are increasingly competitive with silicon solar cells in terms of cost and efficiency. The fabrication method plays an important role in the improvement of these solar cells, since they are responsible for the quality of fabricated thin film and therefore their properties and thus the efficiency of the whole solar cells. Furthermore, the simplicity and low-temperature process make these solar cells more efficient with low cost compared to the other technologies.

Chapter 3
Perovskite-based tandem solar cells

1 Introduction

The combination of the two sub-cells having complementary absorption regions led to a tandem solar cell with higher efficiencies. Since 2014, perovskite solar cells were found to be suitable as top cells over the other solar cell technologies, thanks to their high efficiencies reaching 25.2% [5], combined with their low fabrication cost [78]. In this chapter, two kinds of tandem solar cells will be presented, namely, perovskite/perovskite tandem solar cells and perovskite/CIGS (copper indium gallium di-selenide) tandem solar cells. Before detailing these tandem solar cells and their performances and progress, an overview from the fundamentals to the perovskite based tandem solar cells will be presented first.

2 The definition and principle of tandem solar cells

A tandem solar cell consists of two sub-cells with their absorbers possessing complementary absorption ranges and stacked on top of one another. Each of these sub-cells absorbs a different part of the solar spectrum, thus enabling to exploit a greater range of the solar radiation, leading to a significant increase in conversion efficiency [7, 28]. Figure 27 shows pictures of the perovskite/silicon heterojunction tandem solar cells.

When the tandem solar cell is exposed to light, the top solar cell that has the widest bandgap (between 1.5 and 1.8 eV) absorbs and converts high-energy photons, leaving photons of lower energy to be harvested by the bottom cell having the smaller bandgap (between 0.9 and 1.2 eV) [27, 79]. This operation principle of the tandem solar cell is illustrated in Figure 28.

3 Tandem solar cell configurations

Tandem solar cells can be fabricated mainly in the four terminals or in the monolithic configurations [29, 80].

3.1 Four-terminal tandem

The four-terminal configuration consists of stacking the two sub-cells, fabricated independently, one over the other, as illustrated in Figure 29 [80, 81]. This makes the

https://doi.org/10.1515/9783110760613-004

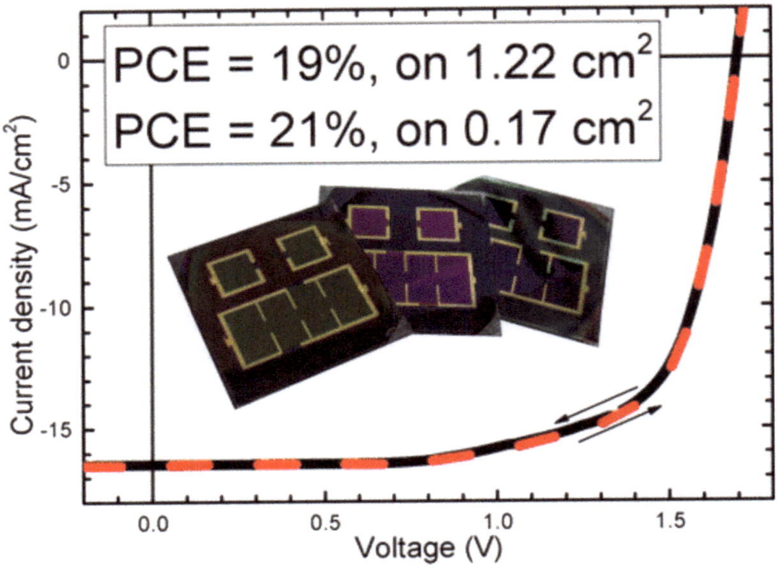

Figure 27: Image of perovskite/silicon heterojunction tandem solar cells reaching efficiencies up to 21.2% and 19.2% for cell areas of 0.17 and 1.22 cm^2, respectively.
Source: Reprinted with permission from WERNER, Jérémie, WENG, Ching-Hsun, WALTER, Arnaud, et al. Efficient monolithic perovskite/silicon tandem solar cell with cell area> 1 cm^2. The journal of physical chemistry letters, 2016, vol. 7, no 1, p. 161–166. Copyright 2016, American Chemical Society.

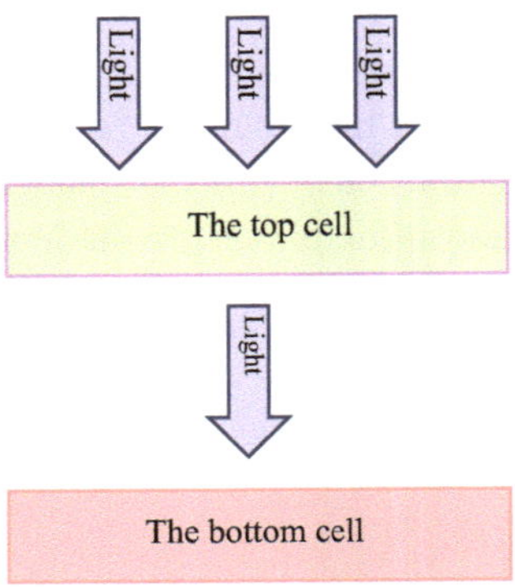

Figure 28: Principle of the tandem solar cell.

fabrication of the four-terminal tandems less demanding than the two-terminal ones. The sub-cells can be connected in series or in parallel.

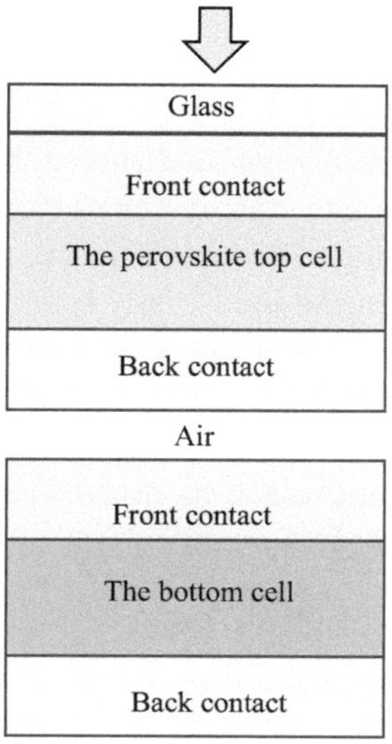

Figure 29: Scheme of the four-terminal tandem solar cell.

3.2 Monolithic tandem solar cell

The monolithic tandem, also called two-terminal configuration, is formed of one single block, as shown in Figure 30. The top solar cell is directly deposited on the bottom one, which requires fewer transparent conductive electrodes and makes them easier to be integrated in a module [29, 80]. Thus, the parasitic absorption due to these electrodes is avoided and the fabrication cost becomes low, thanks to the reduction of the manufacturing steps [29].

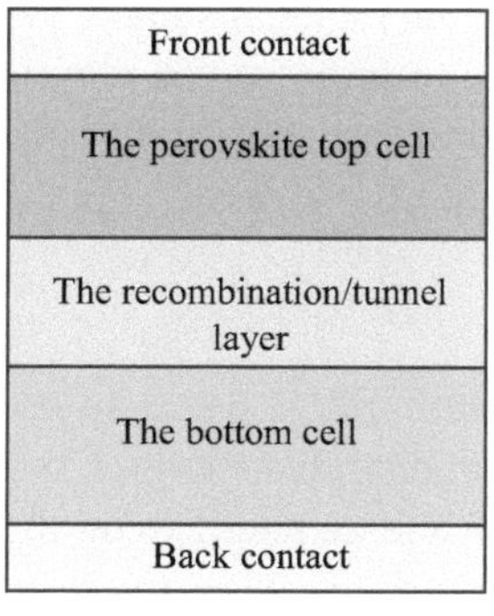

Figure 30: Scheme of the monolithic tandem solar cell.

The fabrication of the monolithic tandem solar cell requires the sub-cells to meet certain criteria, namely [29]:

1. Both sub-cells should be optimized to produce the same current at the point of maximum power.

2. The perovskite top cell must be deposited at a low temperature specially if the bottom cell is sensitive to temperature.
3. The opaque back contact used usually in the perovskite solar cells is unfortunately not suitable for the perovskite sub-cells used in monolithic tandem. Consequently, high transparent contact layer is recommended to replace the ordinary one so that the sub-bandgap light can reach the bottom sub-cell leading to higher efficiency.
4. The perovskite solar cells are usually much smaller than the industrially established silicon solar cells, which requires an adaptation of the scales of the two sub-cells.

Furthermore, monolithic solar cells often produce lower performance compared to four-terminal solar cells; however, they are easier to integrate in a module [7].

4 Tandem solar cells containing the perovskite sub-cells

The unique and special properties of perovskite absorbers make them attractive to be used as a top sub-cell for tandem solar cells [27, 80]. Indeed, several couplings have been envisaged, either in monolithic or in four-terminal configuration. The purpose of this chapter is to present the most efficient tandem solar cells, particularly those with perovskite/perovskite and perovskite/CIGS with four-terminal configuration, and monolithic perovskite/heterojunction silicon solar cells.

4.1 The perovskite/perovskite tandem solar cells

The combined electronic properties and the ability to adjust the bandgaps make the perovskite/perovskite tandem solar cells a good alternative to improve the performance of solar cells with a lower and competitive manufacturing cost [82]. In this part, the best perovskite/perovskite tandem solar cells are presented.

4.1.1 Structure of the perovskite top cell

Good tandem solar cells require the bottom sub-cells with an optical bandgap between 0.9 and 1.3 eV, and the top sub-cells with a larger bandgap between 1.7 and 1.9 eV [82]. Eperon et al. developed perovskite absorbers with a large bandgap on the order of 1.6 eV, namely $FA_{0.83}Cs_{0.17}Pb\ (I_{0.83}Br_{0.17})_3$, for the top solar cell in the four-terminal tandem configuration, and $FA_{0.83}Cs_{0.17}Pb(I_{0.5}Br_{0.5})_3$, with a bandgap of 1.8 eV for the monolithic tandem one [82].

The sub-cell structure developed is a p-i-n configuration, as illustrated in Figure 31, with indium tin oxide as transparent top contact. Methyl [6,6]-phenyl-C61-butanoate

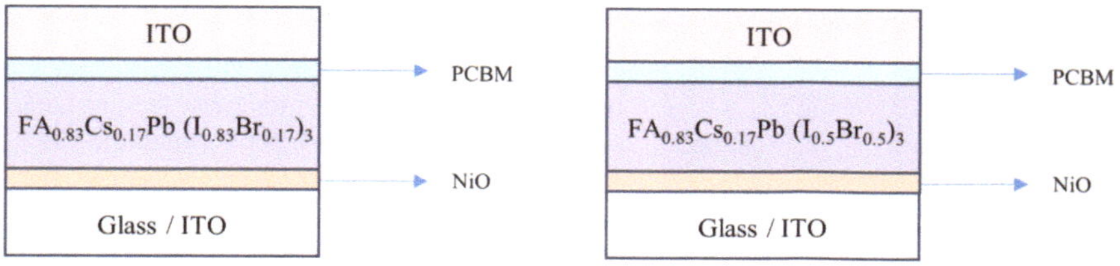

Figure 31: Structure of the perovskite top cell for the four-terminal and the monolithic tandem solar cells, respectively.

was used as the n-type ETL (electron transport layer) and nickel oxide is HTL (hole transport layer).

4.1.2 Structure of the perovskite bottom cell

Eperon et al. developed the perovskite $FA_{0.75}Cs_{0.25}Sn_{0.5}Pb_{0.5}I_3$ absorber with a bandgap of 1.2 eV [82], suitable for use as the bottom cell for the tandem configuration. The structure of the bottom cell is as shown in Figure 32. Poly(3,4-ethylenedioxythiophene):poly (styrenesulfonate) (DEPOT:PSS) was used as the HTL, fullerene (C60) acted as the ETL, bathcuproin was used to improve the collection of electrons, and finally the silver played the role of the metallic contact.

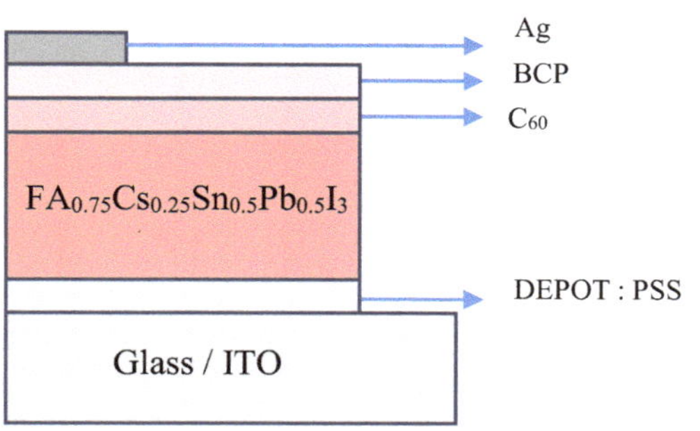

Figure 32: Structure of the perovskite bottom cell.

4.1.3 Tandem solar cell configuration

4.1.3.1 Monolithic tandem solar cell

The coupling of the two sub-cells in monolithic tandem configuration requires a recombination layer, which is in this case the tin dioxide. The scanning electron microscopic (SEM) image shows the configuration of the monolithic tandem solar cell (Figure 33).

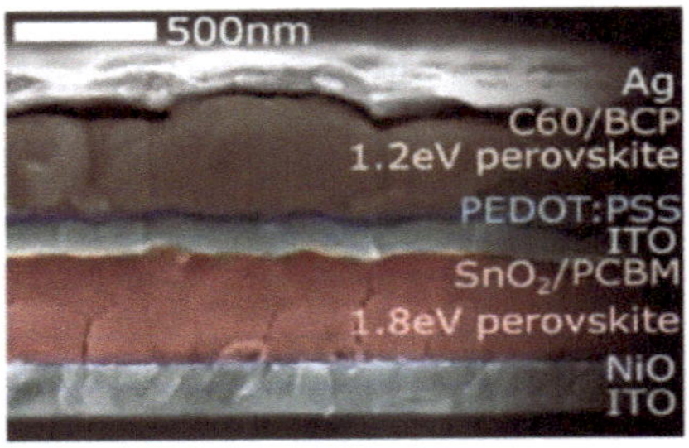

Figure 33: Scanning electron microscopic image of the monolithic perovskite/perovskite tandem solar cell. Source: Reprinted with permission from The American Association for the Advancement of Science, EPERON, Giles E., LEIJTENS, Tomas, BUSH, Kevin A., et al. Perovskite-perovskite tandem photovoltaics with optimized band gaps. Science, 2016, vol. 354, no 6314, p. 861–865.

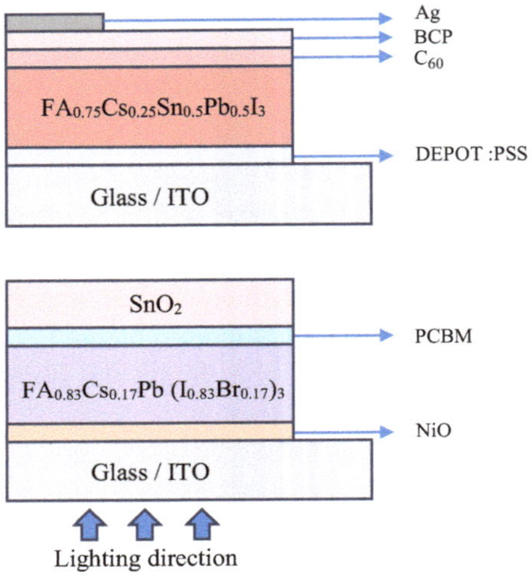

Figure 34: Four-terminal perovskite/perovskite tandem solar cell.

4.1.3.2 Four-terminal tandem solar cell

In the four-terminal configuration, the two sub-cells are mechanically stacked on each other as shown in Figure 34 [82].

4.1.4 Characteristics of tandem solar cells and sub-cells

Measurements of the external quantum efficiency (EQE) for the two configurations of the two tandem solar cells, presented in Figure 35 [82], show that each sub-cell is efficient in a well-defined wavelength range.

Thanks to the complementarity of the efficiency domains of the two sub-cells, the monolithic tandem solar cell achieved a stabilized output of 17%, while the four-terminal configuration reached 20.3% for solar cells with active area of 0.2 cm^2 [82]. More characteristics of the sub-cells and the tandem solar cells are summarized in Table 2 [82].

Moreover, tandem solar cells with the same sub-cells and large surface area, about 1 cm^2, show lower characteristics than those measured for the smallest surface area of 0.2 cm^2, with stabilized power output of 9.5% and 14.8% for the four-terminal and the monolithic tandem solar cells, respectively [82]. These results are grouped in Table 3.

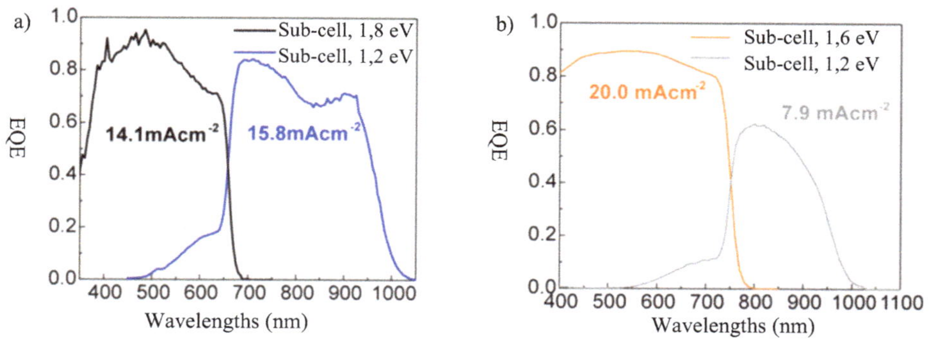

Figure 35: The external quantum efficiency of the two tandem solar cells for (a) monolithic configuration and (b) four-terminal configuration.
Source: Reprinted with permission from EPERON, Giles E., LEIJTENS, Tomas, BUSH, Kevin A., et al. Perovskite-perovskite tandem photovoltaics with optimized band gaps. Science, 2016, vol. 354, no 6314, p. 861–865.

Table 2: Characteristics of the perovskite/perovskite tandem solar cells developed in monolithic and in four-terminal configuration, as well as those of the constituent sub-cells, with an active area of 0.2 cm^2.

solar cells	J_{sc} (mA/cm^2)	V_{oc} (V)	FF	η (%)	Stabilized power output (SPO) (%)
Absorber solar cell of 1.2 eV	26.7	0.74	0.71	14.1	14.8
Absorber solar cell of 1.8 eV	15.1	1.12	0.58	9.8	9.5
Monolithic tandem solar cell	14.5	1.66	0.7	16.9	17
Absorber solar cell of 1.2 eV (four terminal)	7.9	0.74	0.73	4.4	4.5
Absorber solar cell of 1.6 eV (four terminal)	20.3	0.97	0.79	15.7	15.8
Four-terminal tandem solar cell				20.1	20.3

Table 3: Characteristics of tandem solar cells with active area of 1 cm^2.

solar cells	J_{sc} (mA/cm^2)	V_{oc} (V)	FF (%)	η (%)	SPO (%)
Monolithic tandem solar cell	26.7	0.74	0.71	14.1	14.8
Four-terminal tandem solar cell	15.1	1.12	0.58	9.8	9.5

4.1.5 Other perovskite/perovskite tandem solar cells

Great progress has been made in the perovskite/perovskite tandem solar cells, either in monolithic or in four-terminal configurations. Indeed, several researchers have investigated different approaches to further improve this class of solar cells, namely, the perovskite materials, configurations, and composition engineering. Recently, Fang et al. reported on the latest most efficient perovskite/perovskite tandem solar cells, with a record efficiency of 25.6% and 25.4% for monolithic and four-terminal configuration, respectively [83]. Furthermore, Rui Sheng et al. reported higher efficiency of 25.9% for monolithic perovskite/perovskite tandem solar cell using $MAPbBr_3$ as the higher bandgap solar cell and $MAPbI_3$ as the lower one [84].

4.2 Perovskite/CIGS tandem solar cells

4.2.1 Introduction

The Cu(Ga, In)$(Se)_2$, the CIGS, counts among the family of copper indium di-selenide semiconductors of crystallographic structure known as chalcopyrite, which has a direct bandgap between 1 and 3.5 eV [85]. The CIGS has a bandgap between 1.24 and 1.3 eV and is considered to be the best absorber of this family of III–V semiconductors [85].

In 2016, solar cells based on CIGS absorber reached an efficiency of 21.0 ± 0.6% with an open-circuit voltage (V_{oc}) of 0.757 V, a short-circuit current density of 70 mA/cm^2, and a fill factor (FF) of 77.6%, for an area of 1 cm^2 [86]. Recently, the record efficiency of this kind of solar cells achieved 23.4% according to the National Renewable Energy Laboratory chart [5].

The CIGS solar cell has an optimum bandgap for making tandem solar cells [80]. In fact, coupling this type of sub-cell with the perovskite top cell makes it possible to obtain tandem solar cells with improved efficiencies [27]. In this part, a concrete example is exposed, where the structures of the sub-cells, their fabrication methods, and their resulting characteristics are presented without omitting those of the tandem solar cell.

4.2.2 The perovskite sub-cell

4.2.2.1 Structure of the perovskite sub-cell

The perovskite solar cell used in this tandem has the following configuration: FTO (fluorine-doped tin oxide)/c-TiO_2 (compact titanium dioxide)/mp-TiO_2 (mesoporous titanium dioxide):$CH_3NH_3PbI_3$(methylammonium lead iodide)/$CH_3NH_3PbI_3$/spiro-OMeTAD (2,2′7,7′-tetrakis(*N*,*N*-di-*p*-methoxyphenylamine)-9,9′-spirobifluorene)/MoO_3 (molybdenum trioxide)/ ZnO:Al (aluminum-doped zinc oxide)/MgF_2 (magnesium fluoride) [27]. FTO is the substrate, c-TiO_2 and mp-TiO_2 are the electron transporting layers, $CH_3NH_3PbI_3$ is the absorber used, and spiro-OMeTAD is used as a hole transporting layer. Furthermore, the MoO_3 and ZnO:Al form the transparent contact of this perovskite sub-cell, and finally, MgF_2 is the antireflective layer used to improve the performance of this solar cell.

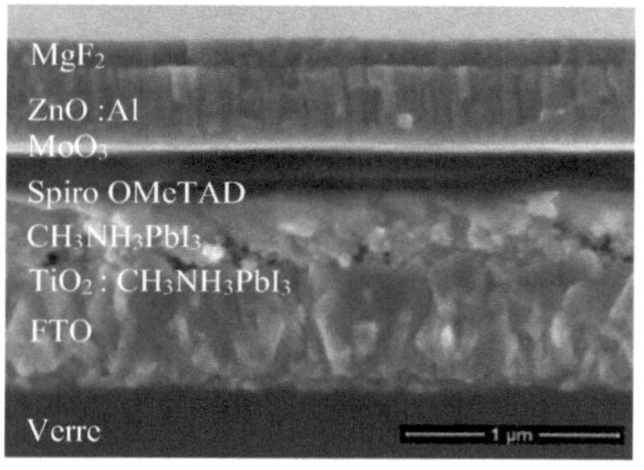

Figure 36: A transverse view of the perovskite solar cell by scanning electron microscope (SEM). Source: Reprinted with permission from KRANZ, Lukas, ABATE, Antonio, FEURER, Thomas, et al. High-efficiency polycrystalline thin film tandem solar cells. The journal of physical chemistry letters, 2015, vol. 6, no 14, p. 2676–2681. Copyright © 2015, American Chemical Society.

The image taken by the SEM is shown in Figure 36. Nevertheless, the layer of compact titanium dioxide (TiO_2) is not visible in this image [27].

4.2.2.2 Fabrication method of the perovskite sub-cell

On an FTO-coated glass substrate, a compact layer of TiO_2 is deposited by spray pyrolysis followed by another mesoporous layer of TiO_2 produced by spin coating [27]. Then, the perovskite is deposited by spin coating and annealed at 90 °C, followed by the elaboration of the spiro-OMeTAD electrode by spin coating also [27]. MoO_3 is deposited by vacuum evaporation, followed by deposition of ZnO:Al by radioactive sputtering [27]. Finally, the metal grid is formed on MgF_2 antireflective coating. Figure 37 shows a schematic description of the various fabrication steps of this perovskite sub-cell.

4.2.2.3 The characteristics of the perovskite sub-cell

The perovskite solar cell usually has an opaque metal contacts; however, the main challenge in the fabrication of the perovskite-based tandem solar cells is to develop PSCs with transparent contacts, so that the photons of energy below the bandgap are transmitted to the bottom cell [27]. In this case, the used perovskite sub-cell, with transparent contact, exhibits the following characteristics, shown in Figure 38.

The transmission curve cited in Figure 38 shows that photons with energy below the bandgap are transmitted with an average of 71%, for the wavelengths between 800 and 1,000 nm [27], and a reflection average of 5% between 350 and 1,100 nm. This low reflection is due to the application of the antireflection layer on the glass substrate and also on the ZnO:Al side [27].

In Table 4, the characteristics of the perovskite solar cell with gold and transparent contacts are compared and presented.

The perovskite reference cell which contains gold contact reached an efficiency of 16.1% compared to 12.1% for the perovskite solar cell with the transparent contact, used as the top cell for perovskite/CIGS tandem configuration. The highest efficiency of the reference cell can be due to the gold contact which reflects the unabsorbed photons to the perovskite solar cell [27]. Furthermore, the perovskite top cell V_{oc} is

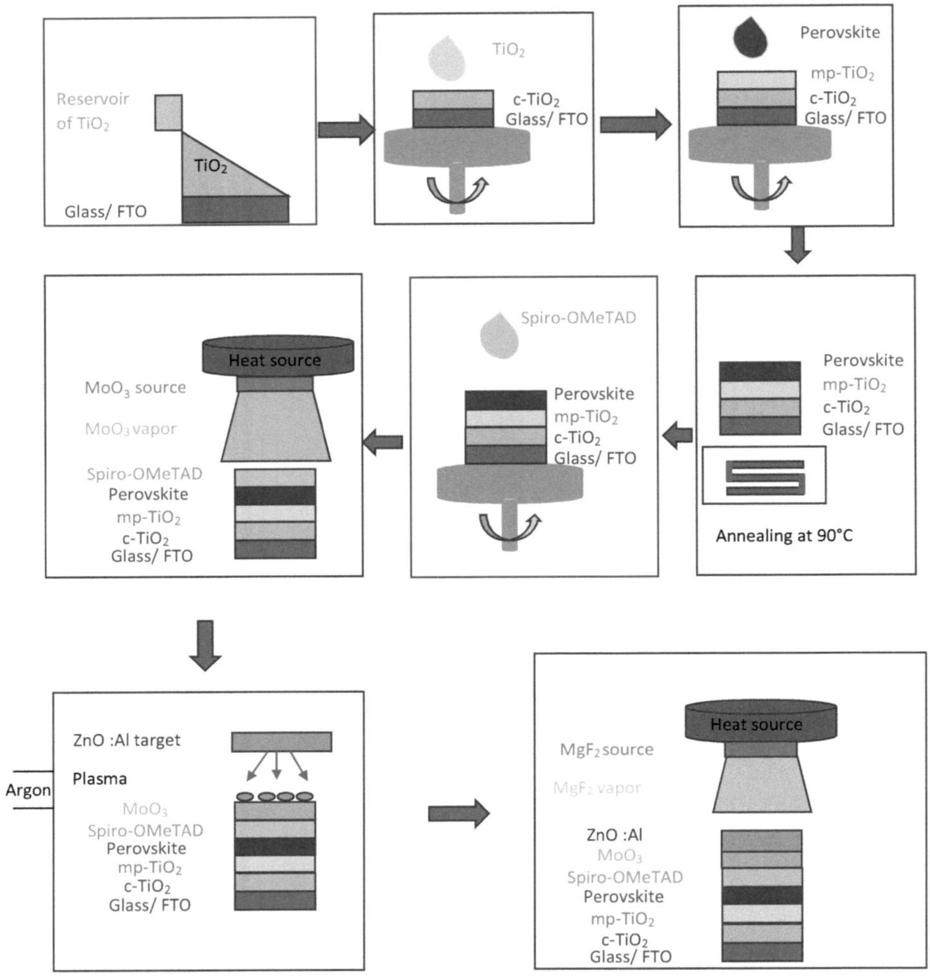

Figure 37: Fabrication method of the perovskite sub-cell.

improved by 26 mV compared to the perovskite reference cell. However, the FF decreased from 75.2% to 70.3%. This reduction is due to the replacement of gold electrode by transparent conductor oxides (TCO) which has a lower conductivity [27]. In addition, the current density decreased by 4.1 mA/cm^2, as a result of the incomplete absorption of the top cell in the wavelength region between 550 and 780 nm [27].

In summary, the performance of the transparent-contact perovskite solar cell is lower than that of gold contact. However, transparency is a necessary condition for developing tandem solar cells.

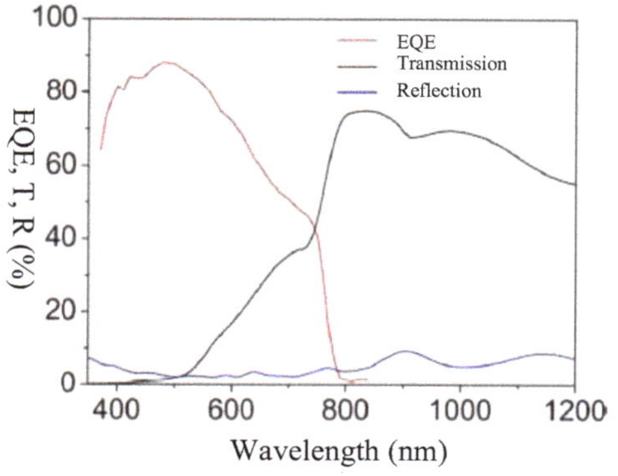

Figure 38: EQE, total transmission T, and reflection R of the perovskite top cell as a function of wavelength.
Source: Reprinted with permission from KRANZ, Lukas, ABATE, Antonio, FEURER, Thomas, et al. High-efficiency polycrystalline thin film tandem solar cells. The journal of physical chemistry letters, 2015, vol. 6, no 14, p. 2676–2681. Copyright © 2015, American Chemical Society.

Table 4: Comparison between the perovskite solar cell with the gold and the transparent-contact.

Solar cell	V_{oc} (mV)	J_{sc} (mA/cm^2)	FF (%)	η (%)	Irradiance (W/cm^2)
The perovskite reference cell (gold contact)	1008	20.8	75.2	16.1	984
Perovskite top cell (transparent contact)	1034	16.7	70.3	12.1	1000

4.2.3 The CIGS sub-cell

4.2.3.1 Structure of the CIGS sub-cell

The CIGS sub-cell is prepared by the co-evaporation technique and consists of the following stacks: substrate/Mo (molybdenum)/CIGS/CdS (cadmium sulfide)/zinc oxide (ZnO)/ZnO:Al, as shown in Figure 39 [27].

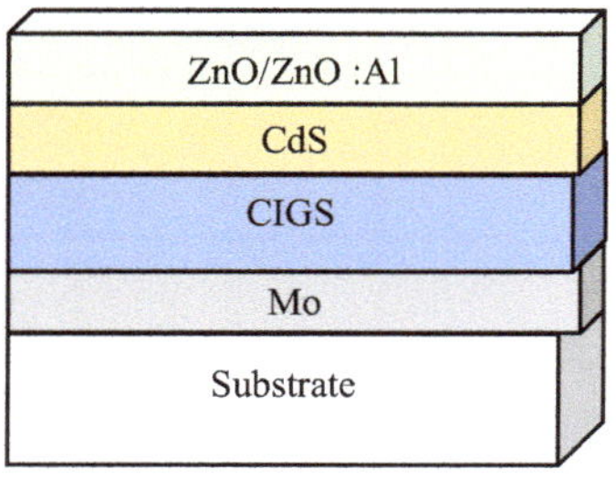

Figure 39: The configuration of the CIGS sub-cell.

4.2.3.2 Method of elaboration of the CIGS sub-cell

On a soda–lime glass, the rear contact consisting of Mo is deposited by magnetron sputtering, and the CIGS layer is prepared by the co-evaporation technique at a temperature of 450 °C and at a pressure (*P*) of the order of 10^{-8} hPa [87]. Then the layer of CdS is deposited via the chemical bath deposition technique, followed by intrinsic ZnO and ZnO:Al layers, deposited by sputtering [27]. Finally, MgF_2 is deposited by electron beam evaporation, in order to reduce reflection losses [27]. Figure 40 shows a schematic description of the various fabrication stages of this CIGS sub-cell.

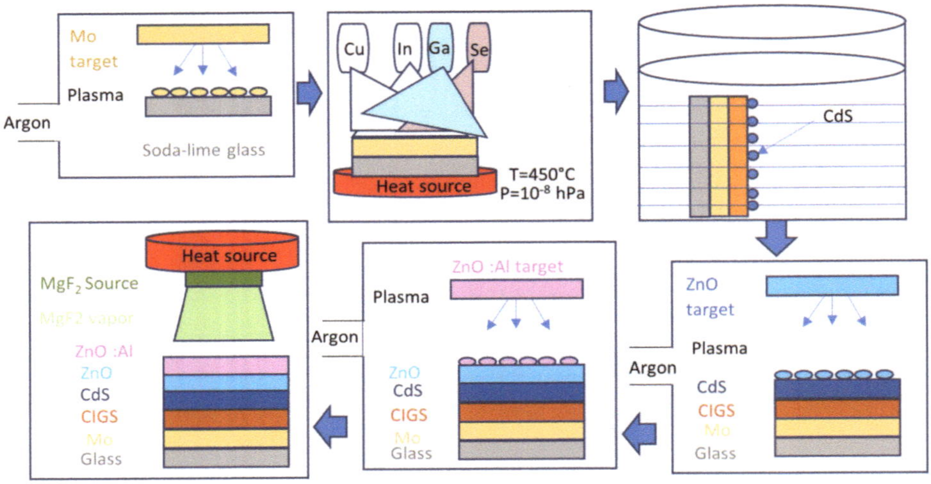

Figure 40: Method of elaboration of the CIGS sub-cell.

4.2.3.3 The characteristics of the CIGS sub-cell

The performances of the stand-alone CIGS solar cell are higher than when used as a bottom cell in a tandem configuration. The characteristics of both CIGS solar cells are presented in Table 5.

Table 5: The characteristics of the stand-alone CIGS solar cell when it is used in tandem configuration.

Solar cell	V_{oc} (mV)	J_{sc} (mA/cm^2)	FF (%)	η (%)	Irradiance (W/cm^2)
The CIGS solar cell (standalone)	689	34.1	78.4	18.4	1000
The CIGS sub-cell (tandem)	661	14.4	77.4	7.4	1000

In fact, this table shows that the CIGS solar cell alone has an 18.4% power conversion efficiency. However, this efficiency decreases to 7.4% when it is below the perovskite sub-cell, with an opening area of 0.213 cm^2. This can be caused by the low illumination intensity, which induces a significant drop in current density as well as a slight decrease in V_{oc} [27].

4.2.4 The tandem solar cell

4.2.4.1 The structure of the perovskite/CIGS tandem solar cell

The four-terminal perovskite/CIGS tandem solar cell consists of the perovskite top cell coupled with the CIGS bottom one [27]. Unlike the CIGS bottom cell which is made in substrate configuration, the perovskite sub-cells has the superstrate structure where the layers of the perovskite sub-cell are made starting from the lowest layer to the last one at the top [27]. The resulting tandem solar cell is schematized as follows in Figure 41.

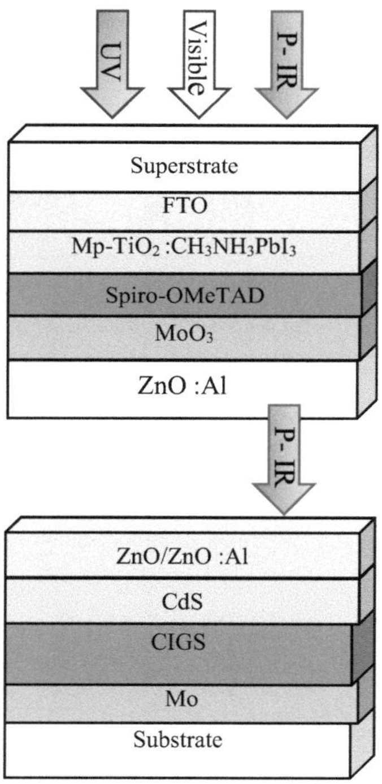

Figure 41: The four-terminal perovskite/CIGS tandem solar cell.

4.2.4.2 The characteristics of the perovskite/CIGS tandem solar cell

The EQE for the sub-cells proves that each solar cell is effective in a range of wavelengths, which is required for the realization of tandem solar cells. The sum of the two EQEs is equal to the EQE of the tandem solar cell, as shown in Figure 42 [27].

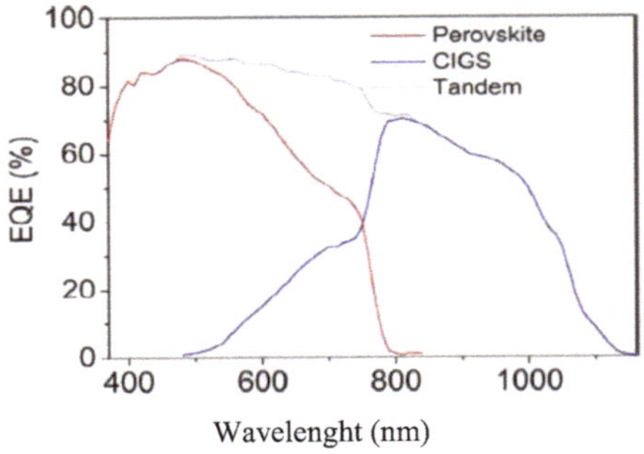

Figure 42: The external quantum efficiency of the sub-cells as well as the tandem solar cell.
Source: Reprinted with permission from KRANZ, Lukas, ABATE, Antonio, FEURER, Thomas, et al. High-efficiency polycrystalline thin film tandem solar cells. The journal of physical chemistry letters, 2015, vol.6, n0 14, p. 2676–2681. Copyright © 2015, American Chemical Society.

The EQE of the tandem solar cell reaches 90% in the wavelength region around 500 nm and decreases for the longer wavelengths. This decrease is mainly due to the optical loss due to the absorption of TCO contained in the tandem solar cell [27].

Table 6: The performances and efficiencies of the sub-cells, as well as the efficiency of the four-terminal tandem solar cell.

Solar cell	V_{oc} (mV)	J_{sc} (mA/cm^2)	FF (%)	η (%)	Irradiance (W/cm^2)
The perovskite top cell (transparent contact)	1034	16.7	70.3	12.1	1000
CIGS bottom cell (tandem)	661	14.4	77.4	7.4	1000
4-terminal tandem solar cell				19.5	1000

Moreover, the efficiency of the resulting four-terminal tandem solar cell reached 19.5%, which is the sum of the two efficiencies of the two sub-cells, as shown in Table 6 [27, 80].

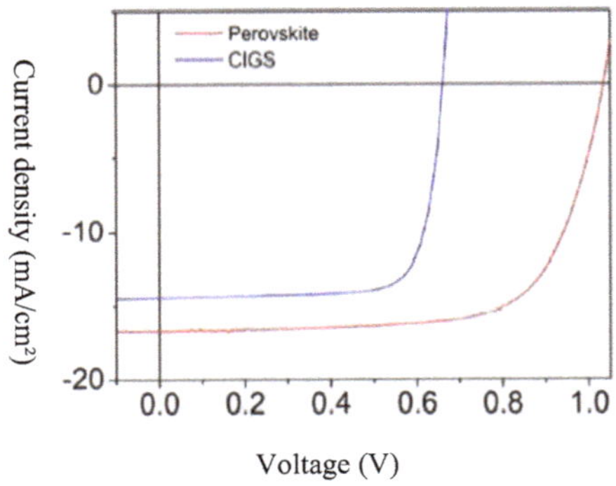

Figure 43: J–V measurements of the two sub-cells.
Source: Reprinted with permission from KRANZ, Lukas, ABATE, Antonio, FEURER, Thomas, et al. High-efficiency polycrystalline thin film tandem solar cells. The journal of physical chemistry letters, 2015, vol. 6, no 14, p. 2676–2681. Copyright © 2015, American Chemical Society.

In addition, the current density versus voltage (J–V) measurements of the sub-cells show that the short-circuit current density (J_{sc}) of the two sub-cells is similar, as shown in Figure 43, which is an important criterion for developing monolithic solar cells [27].

4.2.5 Other perovskite/CIGS tandem solar cells

Recently, IMEC reported a new record efficiency for the perovskite/CIGS tandem solar cells up to 24.6% [88]. This family of solar cells is therefore an interesting alternative to improve CIGS solar cells. In addition, it offers great potential for integration in many industrial applications with high flexibility in terms of size and shapes, since both sub-cells are made of thin films.

5 Conclusion

The integration of perovskite solar cells as top cells, either in perovskite/perovskite tandem solar cells or in perovskite/CIGS tandem solar cells, demonstrates a significant improvement in the efficiency of both bottom cells. Indeed, higher efficiencies were achieved up to 25.6% and 24.6% for perovskite/perovskite tandem solar cells and for perovskite/CIGS tandem solar cells, respectively. However, several requirements must be taken into consideration while designing the perovskite top cells to make them suitable to be used on top of the CIGS and perovskite bottom cells, especially the selection of the materials constituting the perovskite solar cells.

Chapter 4
Perovskite/silicon heterojunction tandem solar cells

1 Introduction

In the last few years, crystalline silicon (c-Si) solar cell efficiency has seen only a slight increase. Actually, they achieved a record efficiency of 26.1% for the homojunction-based solar cells and 26.7% for the heterojunction-based solar cells, certified by the National Renewable Energy Laboratory (NREL) [5, 29]. These efficiencies are close to the maximum theoretical efficiency of crystalline Si-based solar cells of around 29.4% [29]. Moreover, in order to further improve the efficiency of this family of solar cells and to exceed their theoretical limit while keeping a competitive cost, the silicon-based tandem solar cells are considered as the simplest way to reach this objective. This consists of coupling silicon sub-cell with other sub-cells based on low cost and wider bandgap absorbers [29].

Indeed, several solar cells have been used on top of the silicon solar cell, especially the so-called III–V semiconductors. However, their fabrication is expensive and difficult for large areas [29]. Perovskite solar cells offer promising alternatives since they satisfy the whole requirements needed for top cells in tandem with the silicon solar cells, in addition to their continuous efficiency progress of up to 25.2% [5, 29].

Indeed, Loper et al. reported an efficiency of 13.4% for four-terminal perovskite/silicon tandem solar cell, where 6.2% is the efficiency of the perovskite top cell and 7.2% is for the c-Si heterojunction solar cell [89]. The most efficient perovskite/silicon heterojunction tandem solar cells will be detailed in this chapter. Before developing them, an overview of silicon heterojunction solar cells is presented in the following section.

2 The silicon heterojunction solar cell

2.1 Description of the silicon heterojunction solar cell

The silicon heterojunction solar cell (HIT) contains intrinsic silicon layers, and other doped hydrogenated amorphous silicon (a-Si:H), deposited on a c-Si wafer, as illustrated in Figure 44 [90]. The intrinsic layers passivate the c-Si wafer surfaces, which makes it possible to have open-circuit voltage of 750 mV higher than that obtained for the homojunction which does not exceed 700 mV [91]. The p-type a-Si:H forms a heterojunction with the n-type c-Si wafer.

https://doi.org/10.1515/9783110760613-005

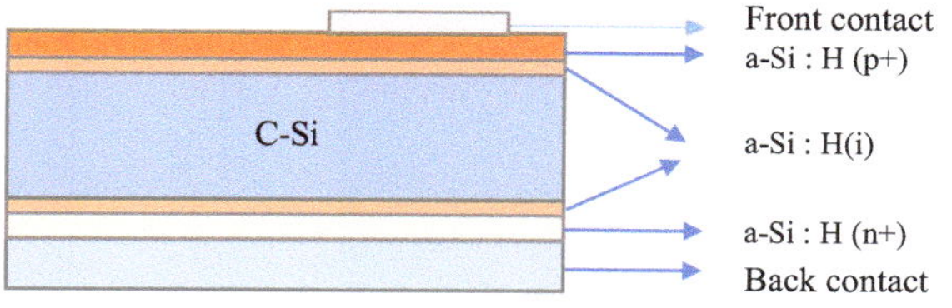

Figure 44: Structure of the silicon heterojunction cell.

2.2 Evolution of the performances of silicon heterojunction solar cells

The efficiency of the HIT solar cell is growing from one year to the next: it evolved from 23.7% in 2011 [90], to 24.7% in 2013 with a standard configuration [91], to 25.6% in 2014 [90], and to 26.3% in March 2017 [92] with an interdigitated back contact. Afterward, an efficiency of 26.6% was obtained and certified by NREL in 2018 [92]. Today, the best HIT solar cell achieved 26.7% efficiency [93]. The evolution of efficiency of the best HIT solar cells, from 2009 to 2020, is represented in Figure 45.

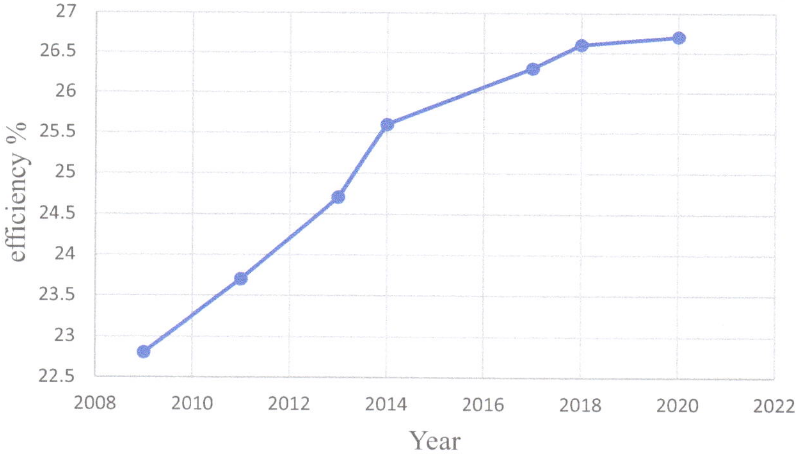

Figure 45: The evolution of the efficiency and the current density of the HIT solar cells.

2.3 The configuration of the record silicon heterojunction solar cell

This section presents the configuration of the silicon heterojunction solar cell which has the best efficiency, of the order of 26.3%, certified by NREL in 2017. This configuration is called silicon heterojunction solar cell with interdigitated back contact [92], illustrated in Figure 46.

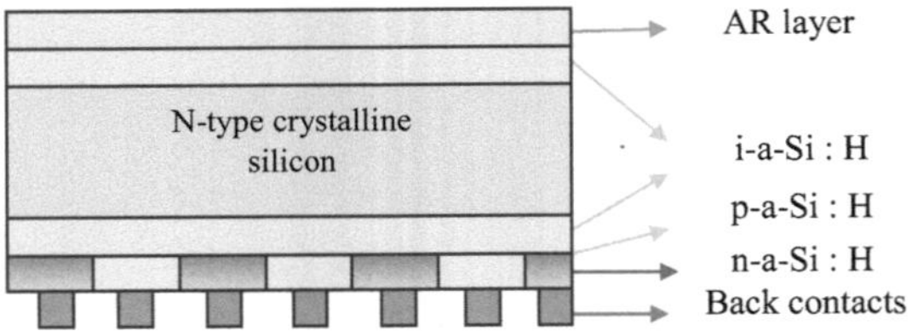

Figure 46: Structure of the record heterojunction solar cell.

The structure of c-Si heterojunction solar cell with interdigitated back contact is developed to reduce optical losses, thus improving the efficiency of the solar cell [90].

2.4 Fabrication method of the record silicon heterojunction solar cell

On the lower side of the n-type Czochralski silicon wafer, an intrinsic amorphous silicon layer and a p-type doped one are deposited, often by plasma-enhanced chemical vapor deposition (PECVD) (see Figure 47) [28, 29, 92] to form a PN heterojunction [90]. On the other side, an intrinsic layer (i) and an n-doped one are deposited to form the back surface field structure [90, 92]. The intrinsic layers of a-Si:H are used to passivate the dangling bonds of the surface of the c-Si wafer [90, 92].

In addition, an antireflective (AR) layer is deposited on the front surface of the c-Si wafer [92]. Finally, back contacts are deposited on the n and p layers of the amorphous silicon [90].

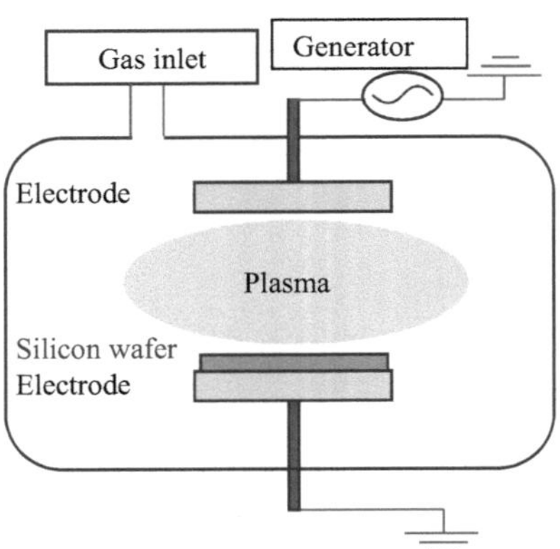

Figure 47: PECVD deposition chamber schemes.

3 Principle of the tandem perovskite/silicon heterojunction solar cell

The perovskite/silicon heterojunction solar cell consists of coupling silicon sub-cell with perovskite sub-cell having wider bandgap than that of the HIT bottom cell [80]. Indeed, perovskite solar cells are transparent to near-infrared radiation [27]. Therefore, as shown in Figure 48, the high-energy photons are absorbed by the perovskite top cell and the infrared photons are transmitted to the silicon bottom cell, thus the tandem solar cell covers a wide spectral range [28].

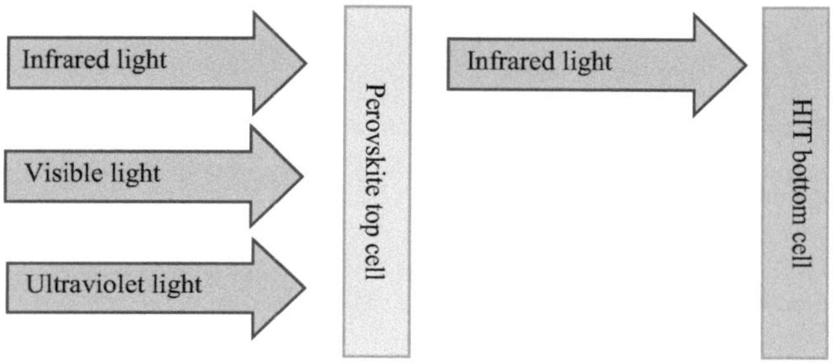

Figure 48: Principle of the tandem perovskite/silicon heterojunction solar cell.

4 Perovskite/silicon heterojunction tandem solar cell

In this part, three monolithic tandem solar cells will be described in detail including their structures, the fabrication process, and their characteristics. The first solar cell yielded 18.1% efficiency [28], 21.2% for the second one [29], and 23.6% for the third solar cell [94].

4.1 The tandem solar cell of 18.1% efficiency

In October 2015, a team from the Helmholtz Center for Materials and Energy in Berlin (HZB) in collaboration with the Swiss Federal Institute of Technology Lausanne (EPFL) developed a tandem solar cell containing perovskite and silicon heterojunction, reaching 18.1% efficiency, with the following structure, as shown in Figure 49 [28]. The configurations and characteristics of the sub-cells and tandem solar cells are detailed further.

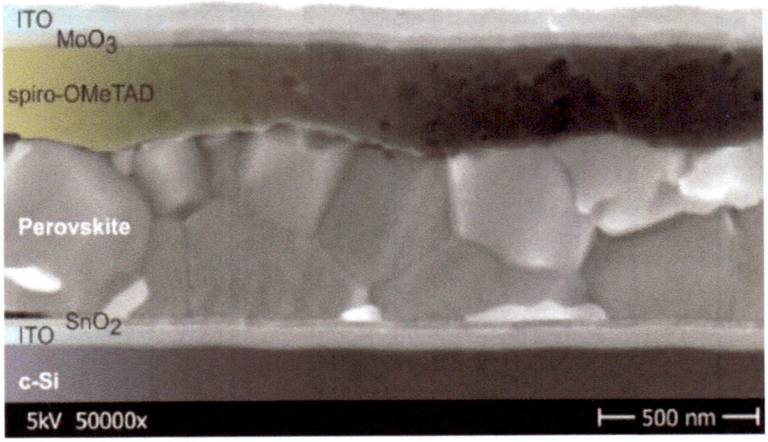

Figure 49: A cross section of the monolithic tandem solar cell seen by SEM.
Source: Reprinted with permission from ALBRECHT, Steve, SALIBA, Michael, BAENA, Juan Pablo Correa, et al. Monolithic perovskite/silicon-heterojunction tandem solar cells processed at low temperature. Energy & Environmental Science, 2016, vol. 9, no 1, p. 81–88.

4.1.1 The perovskite top cell

4.1.1.1 The perovskite top cell structure

The perovskite top cell used in this tandem solar cell consists of the following stacks: tin dioxide (SnO_2)/$(MA)_x(FA)_{1-x}PbI_xBr_{3-x}$/spiro–OMeTAD (2,2′7,7′-tetrakis(*N*, *N*-di-*p*-methoxyphenylamine)-9,9′-spirobifluorene)/MoO_3 (molybdenum trioxide)/ ITO (indium tin oxide) [28], as illustrated in Figure 50.

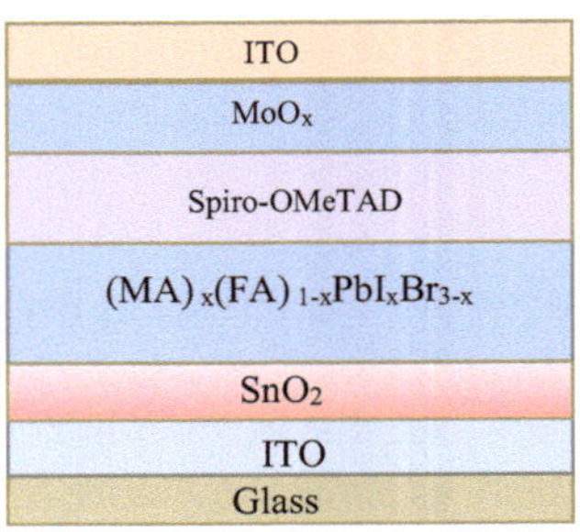

Figure 50: The perovskite top cell structure.

4.1.1.2 Fabrication method of the perovskite top cell

The perovskite top cell is fabricated according to the steps presented in Figure 51. First, SnO_2 is deposited on the front face of the bottom cell by atomic layer deposition (ALD) technique.

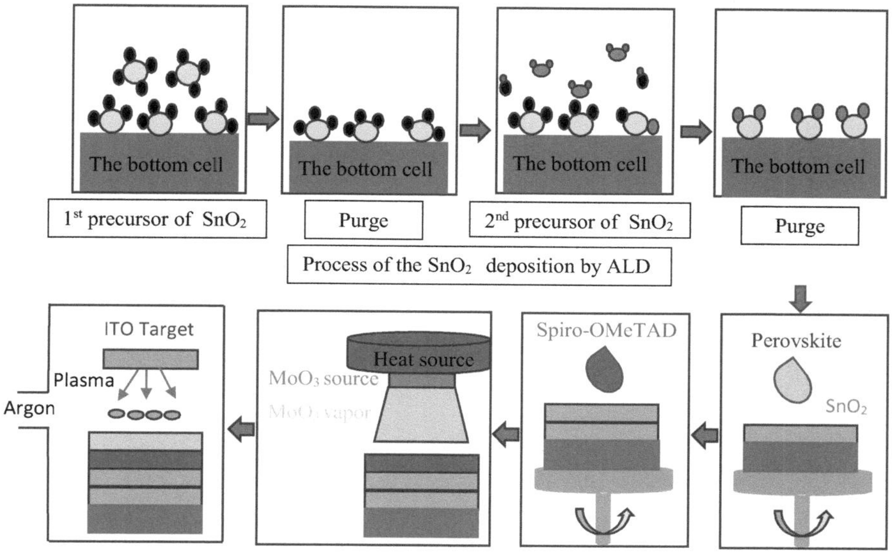

Figure 51: Fabrication method of the perovskite top cell.

Perovskite and spiro-OMeTAD layers are deposited by spin coating in one step, while MoO_3 is deposited by the thermal evaporation technique. Finally, the ITO is sputtered through a mask so that the two ITO layers are aligned with each other [28].

4.1.2 The silicon heterojunction bottom cell

4.1.2.1 The silicon heterojunction bottom cell structure

The silicon heterojunction bottom cell used in this tandem solar cell is of standard configuration [28], as illustrated in Figure 52.

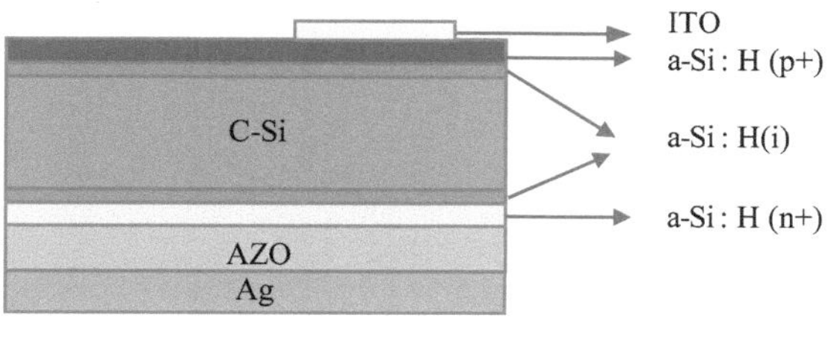

Figure 52: The silicon heterojunction bottom cell structure.

4.1.2.2 Fabrication method of the silicon heterojunction bottom cell

The intrinsic and doped layers of amorphous silicon are deposited by the PECVD on either sides of a polished float-zone n-type silicon wafer [28]. Then, ITO is deposited on the front face by magnetron sputter deposition using a mask to delimit the deposition surface. Furthermore, aluminum-doped zinc oxide and silver (Ag) are also deposited by sputtering on the rear face in such a way as to cover the whole surface [28]. The fabrication method of this silicon heterojunction bottom cell is illustrated in Figure 53.

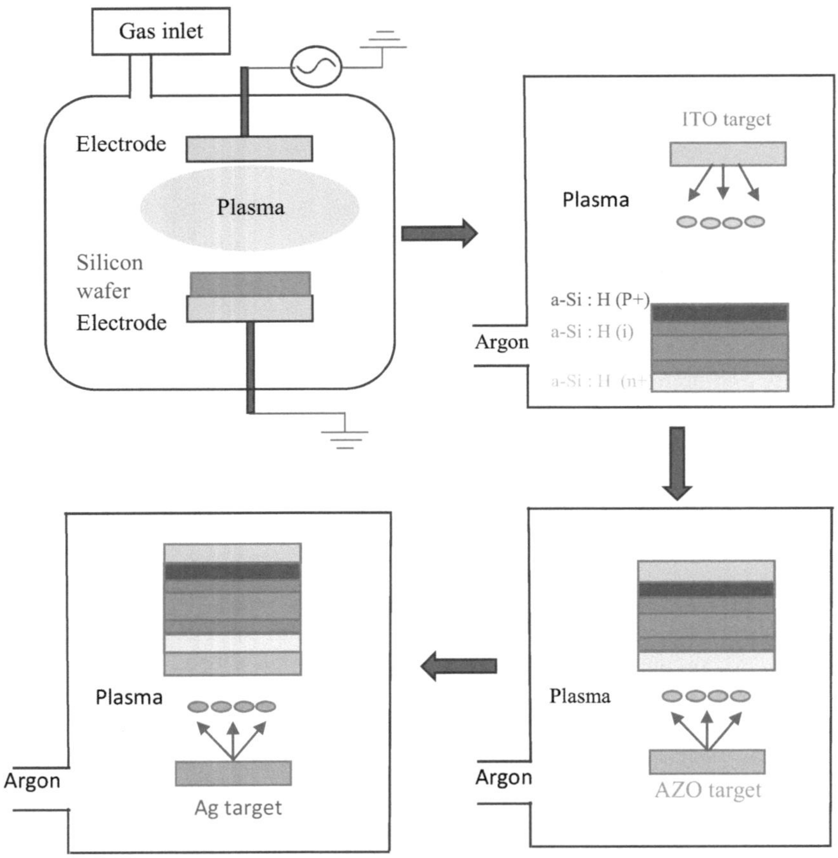

Figure 53: Fabrication method of the silicon heterojunction bottom cell.

4.1.3 Monolithic tandem solar cell

4.1.3.1 Monolithic tandem solar cell structure

The monolithic configuration consists of the perovskite top cell deposited directly on the silicon heterojunction bottom cell, as illustrated in Figure 54.

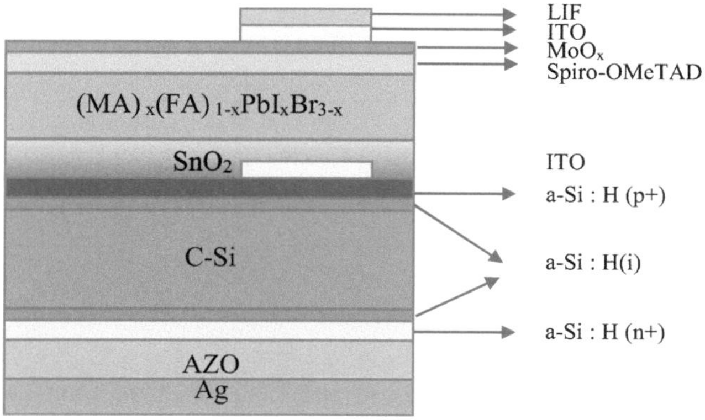

Figure 54: Monolithic tandem solar cell structure.

Lithium fluoride (LIF) is thermally evaporated on the tandem solar cell to reduce reflection losses and consequently improve solar cell performance [28].

4.1.3.2 Monolithic tandem solar cell characteristics

Table 7 shows that the silicon heterojunction bottom cell used in this tandem solar cell exhibits an efficiency of up to 15.7%, while the perovskite top cell achieves 10.4%.

Table 7: The performances of the reference cells.

Reference cells					
solar cell	**Scan**	**J_{sc} (mA/cm^2)**	**V_{oc} (V)**	**FF (%)**	**η (%)**
Silicon		31.3	0.703	71.4	15.7
Perovskite	Reverse	20.1	1.130	68.3	15.5
Perovskite	Direct	20.1	1.048	49.3	10.4

The combination of these two previous sub-cells allowed to achieve an efficiency of up to 19.9%, a short-circuit current density of 14 mA/cm^2, an open-circuit voltage of 1.785 V, and a fill factor (FF) of 79.5%. Nevertheless, the stabilized power output of this tandem solar cell is about 18.1% [28].

Table 8 groups the performances of the best tandem solar cell in direct and reverse scan directions.

This table shows that the tandem solar cell has higher efficiency than those of the reference cells used independently. Therefore, the tandem configuration significantly improves the efficiency of the silicon solar cell with more than 2%. This is considered as an interesting potential for the future of solar cells.

Table 8: The performances of the tandem solar cell.

Tandem solar cell						
solar cell	**Scan**	**J_{sc} (mA/cm²)**	**V_{oc} (V)**	**FF (%)**	**η (%)**	**Stabilized power output**
Tandem	Reverse	14	1.785	79.5%	19.9	18.1%
Tandem	Direct	14	1.759	77.3%	19.1	

4.2 The tandem solar cell of 21.2% efficiency

In 2016, De Wolf et al. reported a record efficiency of 21.2% for perovskite/silicon monolithic tandem solar cell [29]. The configurations and characteristics of the subcells and tandem solar cells are detailed further.

4.2.1 The perovskite top cell

4.2.1.1 The perovskite top cell structure

The perovskite top cell used in this tandem solar cell reached an efficiency of 14.5%. It consists of several stacks, including methylammonium lead iodide ($CH_3NH_3PbI_3$) as absorber. The structure of this perovskite solar cell composed of, IZO (indium zinc oxide)/PEIE (polyethylenimine)/PCBM (phenyl-C61-butyric-acid-methyl-ester)/$CH_3NH_3PbI_3$/spiro-OMeTAD/MoO_x (molybdenum oxide)/IO:H (hydrogenated indium oxide)/ITO, is presented in Figure 55 [29].

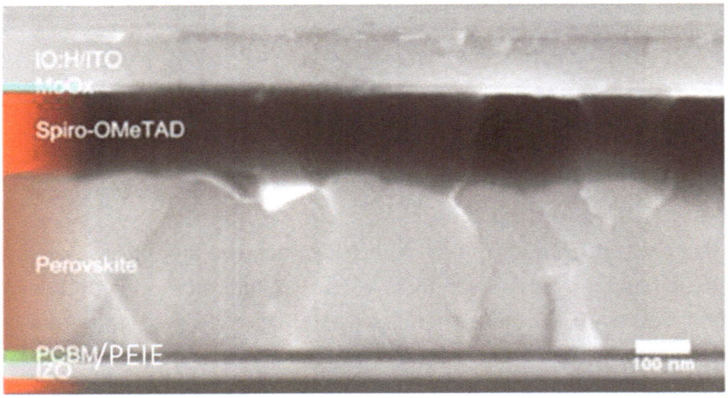

Figure 55: SEM cross section of the perovskite top cell illuminated from MoO_x/IO:H/ITO. Source: Reprinted with permission from WERNER, Jérémie, WENG, Ching-Hsun, WALTER, Arnaud, et al. Efficient monolithic perovskite/silicon tandem solar cell with cell area> 1 cm². The journal of physical chemistry letters, 2016, vol. 7, no 1, p. 161–166. Copyright 2016, American Chemical Society.

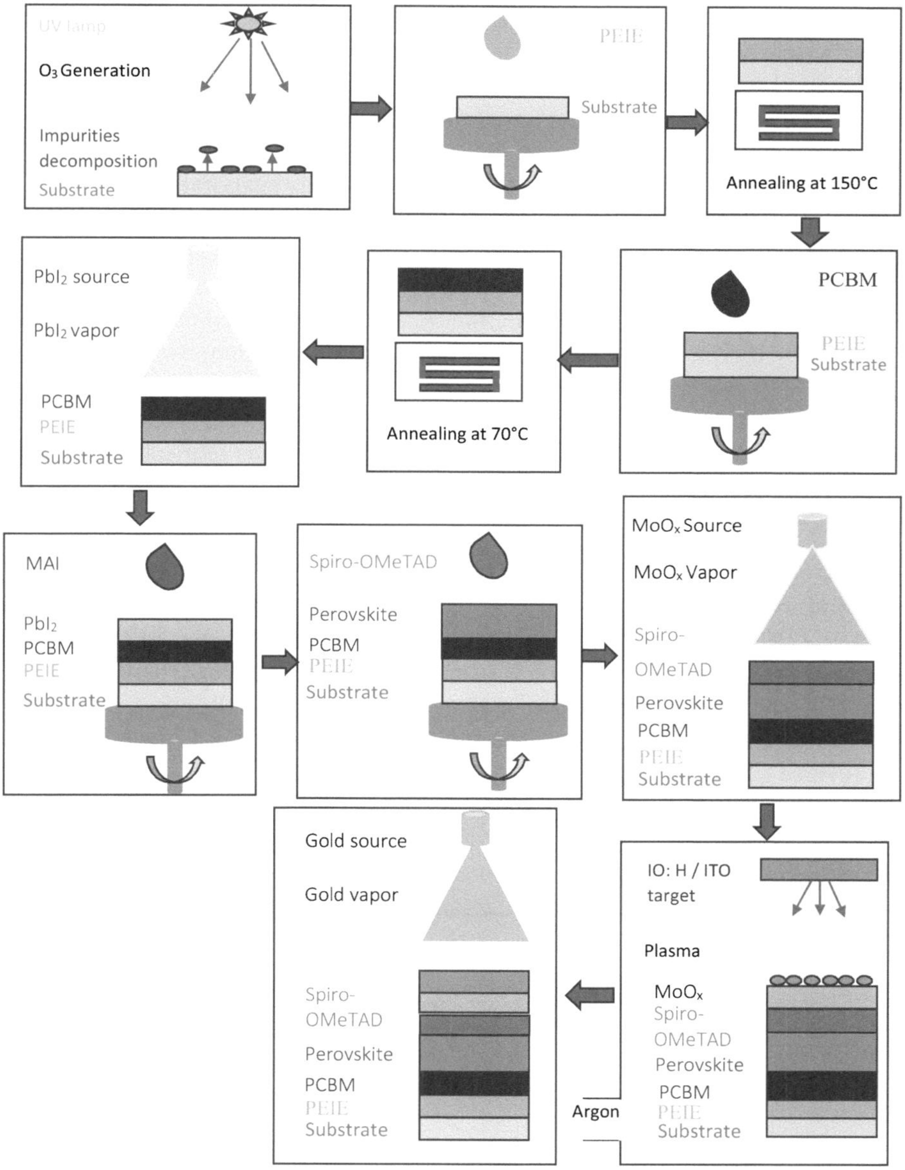

Figure 56: Fabrication method of the perovskite top cell.

4.2.1.2 Fabrication method of the perovskite top cell

The perovskite solar cell used in this tandem was deposited on the intermediate recombination layer according to the following process [29].

A PEIE layer is deposited by spin coater on an UV ozone cleaned substrate and annealed at 150 °C, followed by PCBM layer coated also by spin coating and annealed at 70 °C. Next, lead iodide (PbI_2) was thermally evaporated. Afterward, a solution containing methyl ammonium iodide in isopropanol is deposited by spin coating on PbI_2 to form the perovskite layer. Then the spiro-OMeTAD solubilized in chlorobenzene is deposited by spin coating. Next, a transparent electrode is deposited which is formed of two layers; the first one is composed of 10 nm of MoOx fabricated by thermal evaporation, and the second one is composed of a 110 nm bilayer of IO:H and indium tin oxide IO:H/ITO, deposited by sputtering. Finally, the 100 nm thick front contact is formed by a gold layer by evaporation through a mask. Figure 56 shows a schematic description of the various fabrication stages of this perovskite top cell.

4.2.1.3 Characteristics of the perovskite top cell

The stand-alone perovskite solar cell often uses the glass/ITO combination as substrate [29] and it is illuminated from the glass side. However, in the tandem solar cell, the perovskite sub-cell is illuminated from the MoO_x/IO:H/ITO electrode [29]. The characteristics of the perovskite solar cells illuminated on both sides are presented in Table 9, measured in direct scan; from J_{sc} to V_{oc}, as well as in reverse scan direction.

Table 9: Characteristics of the perovskite solar cells in both illumination directions and in both scan directions.

Perovskite solar cell							
Illuminated from	**The opening surface (cm²)**	**Scan direction**	V_{oc} **(V)**	J_{SC} **(mA/cm²)**	**FF (%)**	η **(%)**	**Maximum power point tracking (MPP)**
Glass/ITO	0.25	Reverse	1.029	17.5	76.9	13.8	14.5
		Direct	1.034	17.5	77.7	14	
Spiro-OMeTAD	0.25	Reverse	1.020	16.2	77.7	12.8	13.1
		Direct	1.020	16.2	79.3	13.1	

This table shows a decrease in perovskite efficiency when illuminated from spiro-OMeTAD. This observation is confirmed by the external quantum efficiency of the illuminated solar cell which exhibits a current loss of 1.5 mA/cm^2, as shown in Figure 57. This can be explained by the parasitic absorption of the spiro-OMeTAD layer [29].

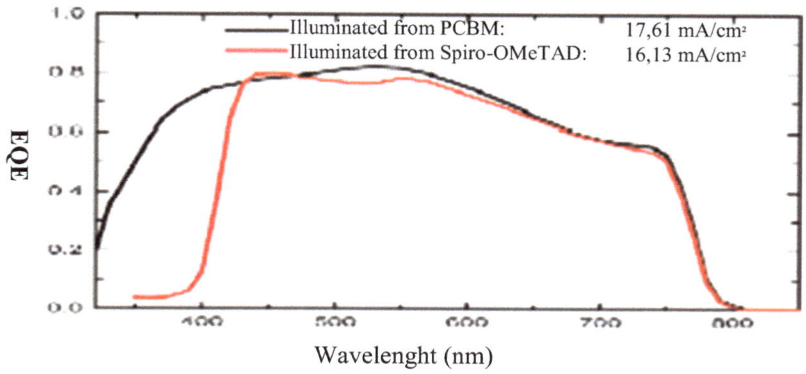

Figure 57: Performances of perovskite solar cell used in single superstrate junction (illuminated from PCBM) and that used in tandem substrate solar cell (illuminated from spiro-OMeTAD). Source: Reprinted with permission from WERNER, Jérémie, WENG, Ching-Hsun, WALTER, Arnaud, et al. Efficient monolithic perovskite/silicon tandem solar cell with cell area> 1 cm^2. The journal of physical chemistry letters, 2016, vol. 7, no 1, p. 161–166. Copyright 2016, American Chemical Society.

Indeed, the spiro-OMeTAD has broad absorption peaks at wavelengths of 380 and 500 nm, which causes the reduction of the photocurrent of the perovskite solar cell illuminated from the spiro-OMeTAD [28].

4.2.2 Silicon heterojunction bottom cell

4.2.2.1 Silicon heterojunction bottom cell structure

The silicon heterojunction bottom cell used in tandem configuration is of standard HIT structure as illustrated in Figure 58 [29].

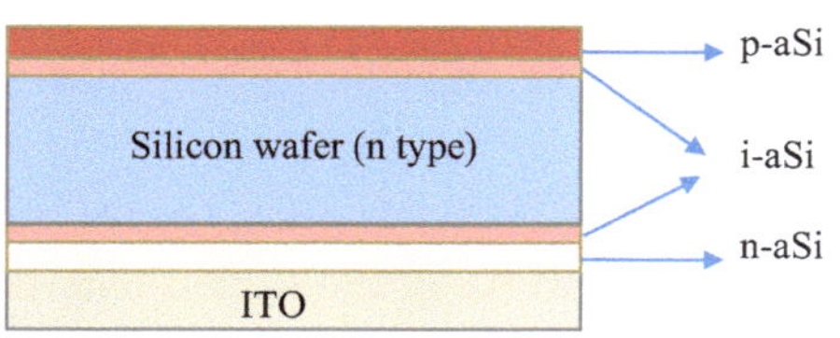

Figure 58: Silicon heterojunction bottom cell structure.

4.2.2.2 Fabrication method of the silicon heterojunction bottom cell

The n-doped float-zone silicon wafer, polished on both sides with a resistivity between 1 and 5 Ω and a thickness between 260 and 300 μm, is dipped in a hydrofluoric acid solution to remove the native oxide on the surface of the wafer. Then intrinsic and doped a-Si:H layers are deposited by the PECVD method, as illustrated in Figure 59 [29].

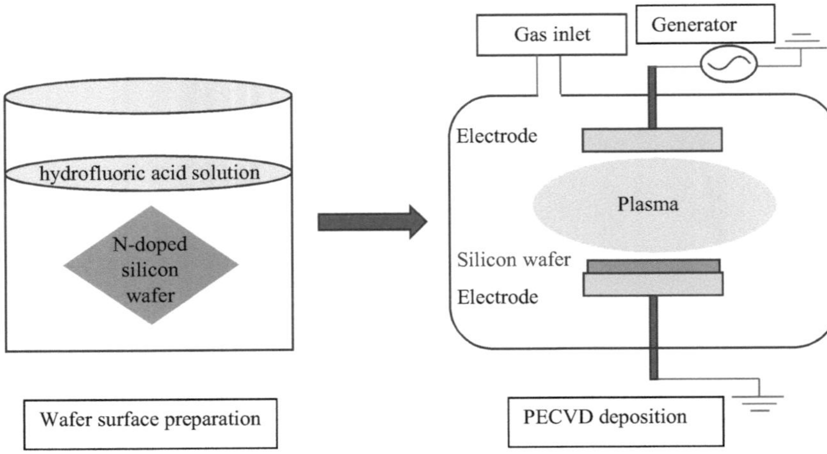

Figure 59: Fabrication method of the silicon heterojunction bottom cell.

4.2.2.3 Characteristics of the silicon heterojunction bottom cell

The heterojunction silicon bottom cell, used in this tandem solar cell, achieved an efficiency of 16.8%. The whole characteristics measured under the standard conditions are presented in Table 10 [29].

Table 10: Characteristics of the silicon heterojunction bottom cell.

Silicon heterojunction bottom cell				
Aperture area (cm^2)	**V_{oc} (V)**	**J_{SC} (mA/cm^2)**	**FF (%)**	**η (%)**
1.22	0.704	32.1	74.4	16.8

4.2.3 The monolithic tandem solar cell

4.2.3.1 Monolithic tandem solar cell structure

The monolithic tandem solar cell studied below achieved 21.1% efficiency [29]. It consists of a perovskite planar top cell, deposited at 150 °C on the silicon heterojunction bottom cell. The IZO) is used as a recombination layer to connect the two sub-cells due to its optimal optical and electrical properties [29]. The recombination layer and the upper electrode are aligned, defining the active area of the tandem solar cell [29]. The detailed structure is shown in Figure 60.

4.2.3.2 Characteristics of the monolithic tandem solar cell

The characteristics of the perovskite/silicon tandem solar cell are summarized in Table 11 [29].

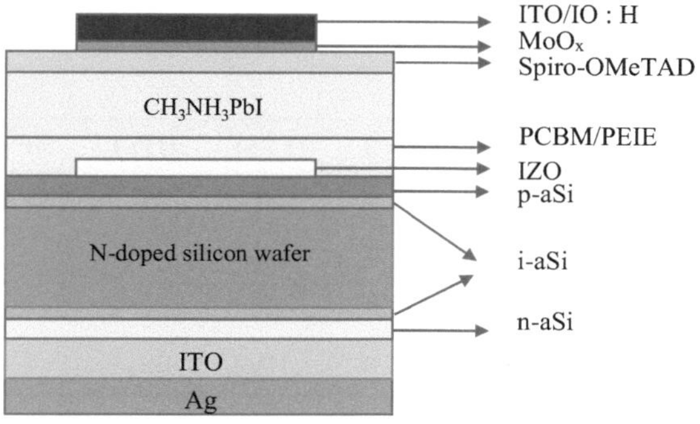

Figure 60: Monolithic perovskite/silicon tandem solar cell structure.

Table 11: The characteristics of the tandem solar cell for both aperture area, without antireflection layer.

Perovskite/silicon tandem solar cell					
Aperture area (cm²)	**Scan direction**	**V_{oc} (V)**	**J_{SC} (mA/cm²)**	**FF (%)**	**η (%)**
1.22	Reverse	1.694	13.4	72.9	16.6
	Direct	1.7	13.4	75.5	17.2
0.17	Reverse	1.668	13.6	76.7	17.4
	Direct	1.67	13.8	78.6	18.1

The combination of the two sub-cells in tandem configuration allowed higher efficiencies than those of the sub-cells used independently. Indeed, they reach 18.1% and 17.2% with an aperture area of 0.17 and 1.22 cm^2, respectively.

4.2.3.3 Effect of the antireflective layer on the performances of the tandem solar cell

Moreover, the addition of an AR layer which aims to reduce the amount of reflected light ensures a significant generation of charge carriers. This has further improved the performance of the tandem solar cell, as summarized in Table 12.

This table demonstrates that adding the AR layer enhances the performance of the tandem solar cells. Therefore, they achieve a record efficiency of 21.2% and of 19.2%, respectively, for an aperture area of 0.17 and 1.22 cm^2.

Furthermore, the measurements of the current density as a function of the voltage (J–V) also confirmed the contribution of the AR layer to the enhancement of the tandem solar cell performance and more precisely of the current density, as shown in Figure 61.

Table 12: The characteristics of the tandem solar cell for both aperture areas, with the antireflective.

Perovskite-silicon tandem solar cell						
Aperture area (cm^2)	**Scan direction**	**V_{oc} (V)**	**J_{SC} (mA/cm^2)**	**FF (%)**	**η (%)**	**MPP**
1.22	Inverted	1.701	16.1	70.1	19.2	19.2
	Direct	1.703	16.1	70.9	19.5	
0,17	Inverted	1.69	15.9	77.6	20.9	21.2
	Direct	1.692	15.8	79.9	21.4	

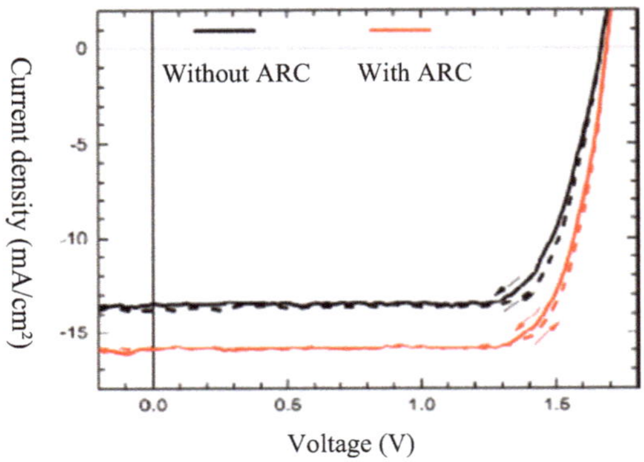

Figure 61: *J–V* measurements for the same tandem solar cell with an aperture area of 0.17 cm^2 with and without antireflective layer.
Source: Reprinted with permission from WERNER, Jérémie, WENG, Ching-Hsun, WALTER, Arnaud, et al. Efficient monolithic perovskite/silicon tandem solar cell with cell area> 1 cm^2. The journal of physical chemistry letters, 2016, vol. 7, no 1, p. 161-166. Copyright 2016, American Chemical Society.

4.3 The tandem solar cell of 23.6% efficiency

In 2017, Kevin A. Bush et al. realized a monolithic perovskite/silicon tandem solar cell with a record efficiency of 23.6% for an area of 1 cm^2 [94]. The configurations and characteristics of the sub-cells and tandem solar cells are detailed further.

4.3.1 The perovskite sub-cell

4.3.1.1 The perovskite sub-cell structure

The perovskite sub-cell used in this tandem configuration consists of the following stacks: glass/ITO/NiO/$Cs_{0.17}FA_{0.83}Pb(Br_{0.17}I_{0.83})_3$/LiF/PCBM/$SnO_2$/ZTO (zinc tin oxide)/ITO/LiF/Ag [94]. NiO_x was used as a hole transport material, and PCBM is the electron transport layer chosen for this solar cell [94]. Moreover, the layers of SnO_2, ZTO, and

ITO are used to facilitate the transport of the electrons and to participate in the improvement of solar cell performances. In addition, LiF was used to reduce reflection losses, and demonstrated an ability to increase the FF. Finally, Ag is the metallic contact used in this configuration presented in Figure 62.

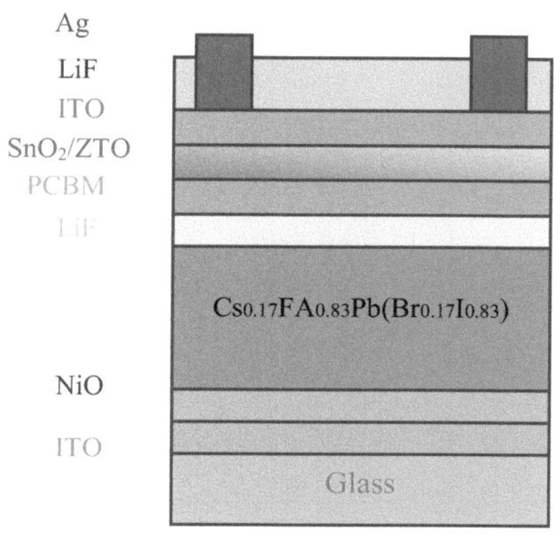

Figure 62: The perovskite sub-cell structure.

4.3.1.2 Fabrication method of the perovskite top cell

The perovskite top cell is deposited according to the following steps [94]:

First, the ITO-coated glass substrates were cleaned with acetone and isopropanol followed by UV ozone treatment. Then, the NiO was deposited by spin coater and annealed at 190 °C for 10 h. This film was rapidly introduced into a dry air box, where the $Cs_{0.17}FA_{0.83}Pb(Br_{0.17}I_{0.83})_3$ perovskite film was formed from a stoichiometric solution, containing cesium iodide, formamidinium iodide, PbI_2, and lead bromide , in a mixture of *N*,*N*-dimethylformamide and dimethylsulfoxide. This stoichiometric solution was spun on the top of NiO and during the centrifugation, chlorobenzene was rapidly added 5 s before the end of the spinning operation, as an anti-solvent to assist the crystallization of perovskite. The films were annealed on a hot plate at 50 °C for 1 min and then at 100 °C for 30 min. Afterward, 1 nm of LiF and 10 nm of PCBM are formed by thermal evaporation. Then, 4 nm of SnO_2 is deposited by the pulsed chemical vapor deposition technique at 100 °C, followed by 2 nm ZTO, which was deposited by combining the ALD processes of SnO_2 and ZnO in a repetitive cycle consisting of three cycles of SnO_2 followed by three cycles of ZnO. Then a layer of 150 nm of ITO is deposited by sputtering. Finally, the Ag electrode is sputtered or evaporated, followed by evaporation of the second AR layer LiF, to reduce the reflection losses and therefore improve the performance of tandem solar cells [28]. Figure 63 shows a schematic description of the various fabrication stages of this perovskite top cell.

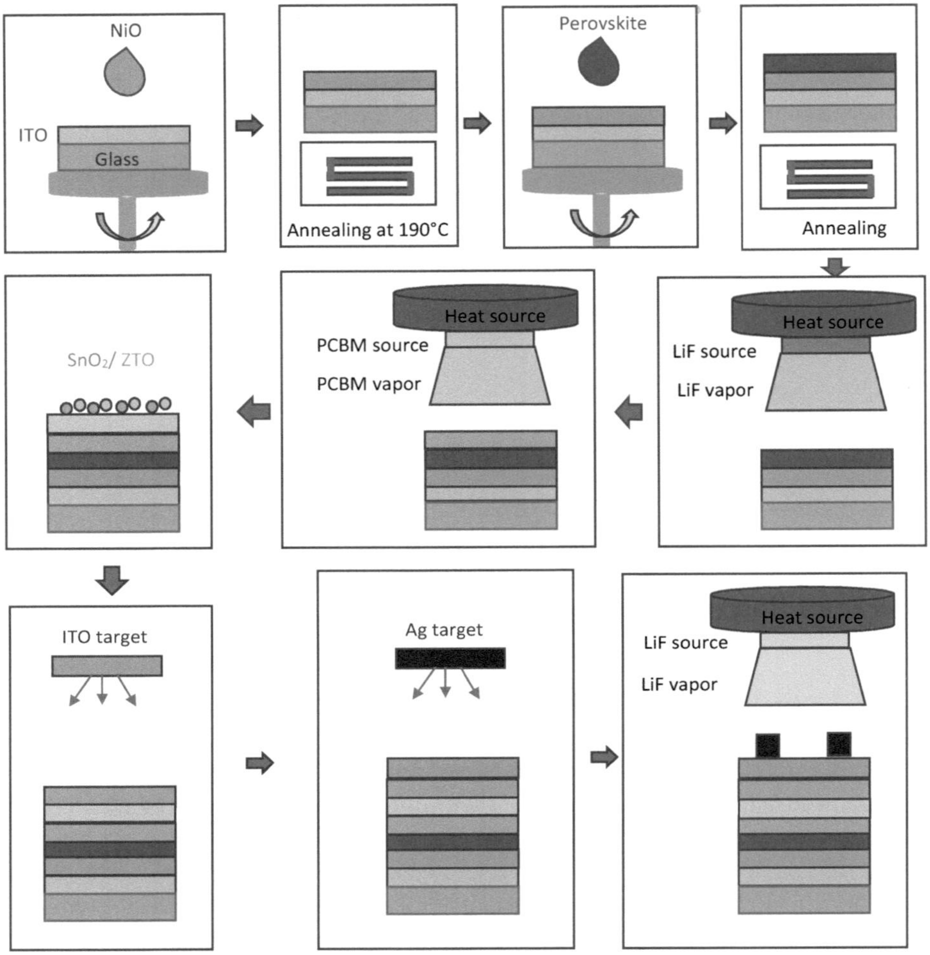

Figure 63: Fabrication method of the perovskite top cell.

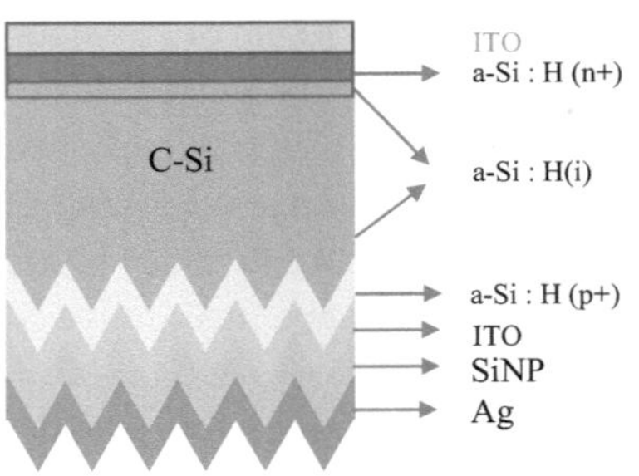

Figure 64: Silicon heterojunction bottom cell structure.

4.3.2 Silicon heterojunction bottom cell

4.3.2.1 Silicon heterojunction bottom cell structure

The silicon heterojunction bottom cell used in this tandem solar cell is illustrated in Figure 64 [94].

4.3.2.2 Fabrication method of the silicon heterojunction bottom cell

A 280-μm-thick float-zone silicon wafer is texturized from its back surface to form random pyramids while the front surface is left flat to deposit perovskite by spin- coating technique [94]. The intrinsic (i) and doped (p) amorphous silicon films are deposited on the textured side by the PECVD technique, with a thickness of 7 and 15 nm, respectively. On the other polished side of the silicon, the intrinsic and doped silicon films (n) were then deposited [94]. Then, the recombination layer between the two sub-cells consists of 20 nm of ITO, sputtered through a mask, defining an area of 11 × 11 mm^2 [94]. On the back side, another layer of ITO is formed of 20 nm thick, followed by another layer formed of 300 nm of silicon nanoparticles (SiNP) deposited by the so-called spray-coating technique through a mask of stainless steel [94]. Finally, an Ag layer of 200 nm is formed by sputtering. The latter forms with SiNP an AR layer [94]. Figure 65 shows a schematic description of the various fabrication stages of this silicon heterojunction bottom cell.

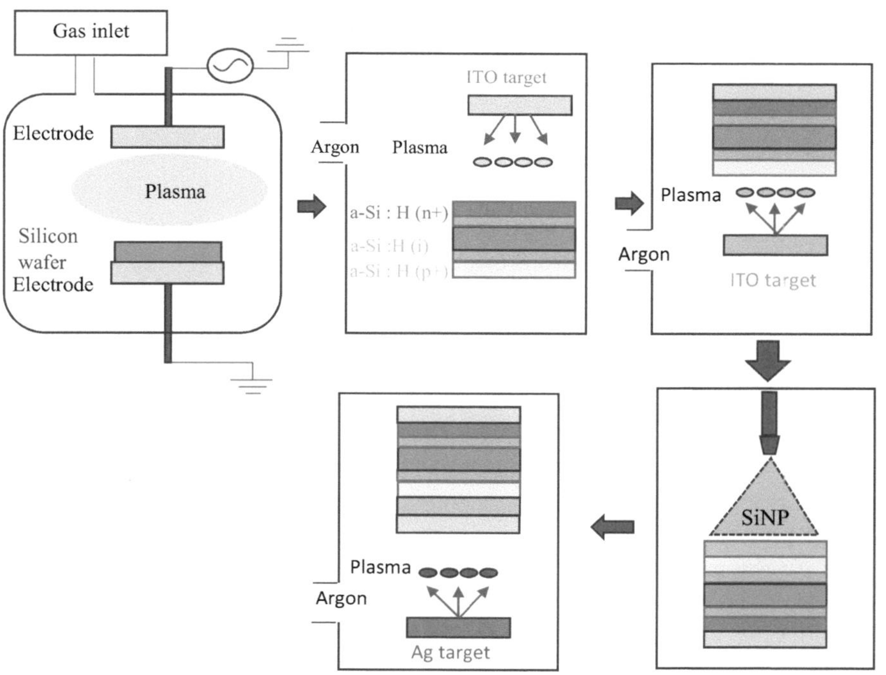

Figure 65: Fabrication method of the silicon heterojunction bottom cell.

4.3.3 The monolithic tandem solar cell

4.3.3.1 Structure of the tandem solar cell

This monolithic configuration consists of the perovskite top cell deposited directly on the silicon heterojunction bottom cell [94], as illustrated in Figure 66.

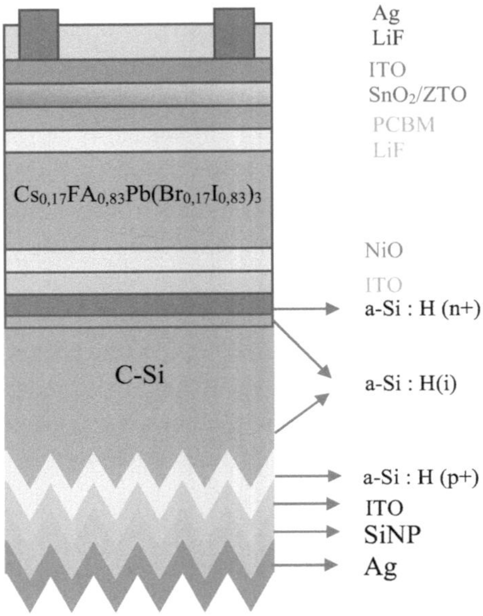

Figure 66: Structure of the monolithic record tandem solar cell.

4.3.3.2 Characteristics of the tandem solar cell

The best combination of these two previous sub-cells has the following parameters certified by NREL: open-circuit voltage of 1.65 V, short-circuit current density of 18.1 mA/cm^2, the FF of 79.0%, and record efficiency of 23.6% [94]. The measurements of current density as a function of voltage are shown in Figure 67.

4.4 Recent progress in perovskite/silicon tandem solar cells

In the last phase of preparation of this book, other efficiency records have been achieved for the perovskite/silicon tandem solar cells. They will be presented in this section.

Indeed, Chen et al. have explored the addition of MACl and MAH_2PO_2 additives to the perovskite precursors in order to increase the grain size of the perovskite layer, resulting in a low defect density and a higher efficiency of the perovskite/

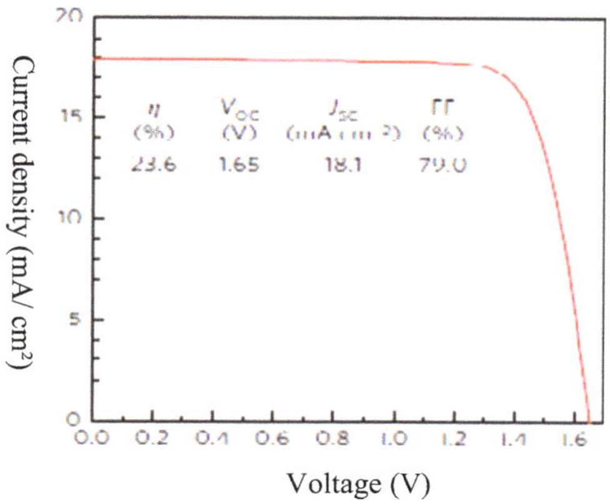

Figure 67: J–V measurements of the perovskite/silicon tandem solar cell.
Source: Reprinted with permission from Springer Nature : Nature Energy, 23.6%-efficient monolithic perovskite/silicon tandem solar cells with improved stability, BUSH, Kevin A., PALMSTROM, Axel F., ZHENGSHAN, J. Yu, et al., 2017, vol. 2, no 4, p. 1–7.

silicon tandem solar cell up to 25.4%, V_{oc} of 1.80 V, J_{sc} of 17.8 mA/cm^2, and FF of 79.4% [95]. This result was reached with a top cell of 1.64 eV bandgap with $Cs_{0,15}(FA_{0,83}\ MA_{0.17})_{0.85}Pb(I_{0,8}Br_{0,2})_3$ perovskite absorber. The structure of the tandem solar cell is presented in Figure 68.

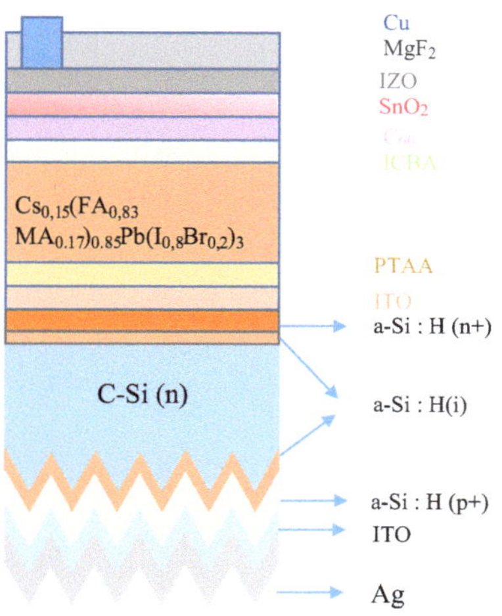

Figure 68: Structure of the monolithic tandem solar cell of 25.4% efficiency.

Furthermore, Xu et al. elaborated a triple halide perovskite with wide bandgap of 1.67 eV and demonstrated stability. The semitransparent perovskite device used in the tandem solar cell enables 80% near-infrared transmittance. Therefore, an efficiency of 27.13% was reached with J_{sc} of 19.12 mA/cm^2, FF of the order of 75.3%, and V_{oc} up to 1.886 V [96].

The V_{oc} achieved in this tandem solar cell is considered as one of the highest V_{oc} reported for perovskite/silicon tandems with an efficiency exceeding 25%. This high efficiency was associated with materials used, their thicknesses, and of course to the structure of the whole solar cell presented in Figure 69.

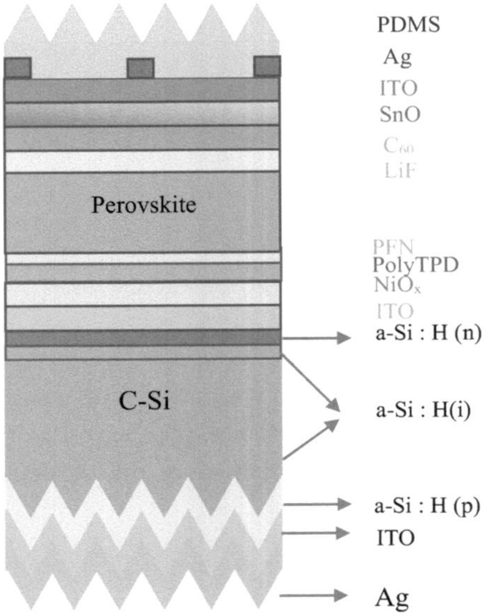

Figure 69: Structure of the monolithic tandem solar cell of 27.13% efficiency.

In 2018, Oxford PV reached two world records' efficiency, for perovskite/silicon tandem solar cell, of 27.3% in June and 28% in December, respectively [97]. Both records were certified by NREL. Recently, in 2020, a perovskite-silicon tandem solar cell with a record efficiency of 29.1% was developed at the Helmholtz Zentrum Berlin, and this achievement was certified by NREL [93]. However, no details about the materials and structures used in these last highest tandem solar cells are available at the time of writing.

4.5 Perovskite/silicon monolithic tandem solar cells evolution

The efficiency of the perovskite/silicon tandem solar cell has been in continuous evolution since its appearance. Researchers are deploying all the means to advance research and innovation for this tandem solar cell category, which is considered to be the future of photovoltaic solar cells. In this section, the evolution of the monolithic perovskite/silicon tandem solar cells is presented in Figure 70.

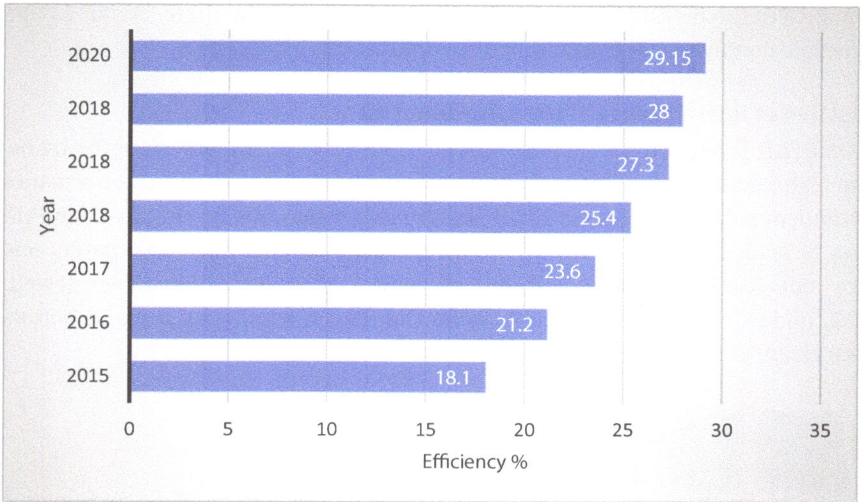

Figure 70: Perovskite/silicon monolithic tandem solar cell evolution.

5 Means to improve the performances of the tandem solar cells

The tandem solar cells reported to date are still far from the theoretical limit, which is of the order of 46% [98]. There is a matrix of parameters to be studied to further improve the tandem solar cells. The main factors influencing their performance are as follows:

- **The use of sub-cells with good performances**

The efficiency of the tandem solar cell can be improved by using sub-cells with record efficiencies but also by shifting the bandgaps of bottom cells to lower values and those of top cells to larger bandgaps [27].

- **Increase the thickness of the perovskite layer**

Current losses can be reduced by increasing the thickness of the perovskite layer, and thus the efficiency of the tandem solar cells can be improved [29].

- **The choice of the nature of transparent contact**

The perovskite top cell requires contact layers with high conductivity and optimum transparency [99]. To further improve the performance of tandem solar cells, it is recommended to optimize the transparent contacts [27, 29], to increase the current of the bottom cell by decreasing the parasitic absorption of the other layers. In fact, many researchers demonstrated that the graphene could be the right alternative for all these reasons, since it combines an excellent optical transmission of the order of 97.4% and a sheet resistance of about 125 Ω/sq [99, 100]. In addition, the electrical performance of the perovskite solar cell containing graphene is quite similar to that of perovskite containing gold as a metallic contact [99].

– Reduction of the thickness of the spiro-OMeTAD

Spiro-OMeTAD is known for its absorption, especially for the wavelengths below 400 nm [29]. Reducing this absorption will be a solution to improve the performance of the tandem solar cell, especially for illuminated solar cells from spiro-OMeTAD. This can be achieved by replacing spiro-OMeTAD with another more transparent material or reducing its thickness [29]. Indeed, several thicknesses have been tested: 200, 130, and 60 nm [29]. The effect of spiro-OMeTAD thickness on top and bottom cell performances is presented in Figure 71.

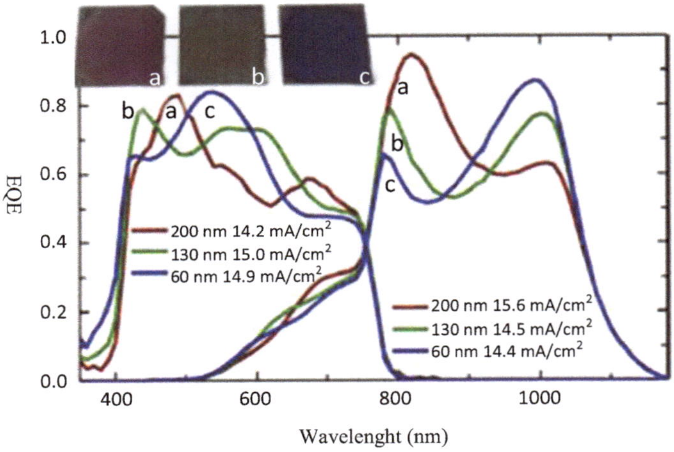

Figure 71: Effect of spiro-OMeTAD thickness on top and bottom cell performances.
Source: Reprinted with permission from WERNER, Jérémie, WENG, Ching-Hsun, WALTER, Arnaud, et al. Efficient monolithic perovskite/silicon tandem solar cell with cell area 1 cm^2. The journal of physical chemistry letters, 2016, vol. 7, no 1, p. 161–166. Copyright 2016, American Chemical Society.

For wavelengths below 400 nm, the current is improved by 0.3 mA/cm^2, which is due to the reduction in the thickness of the spiro-OMeTAD. Furthermore, a simulation confirmed that good results are found with a thickness of 150 nm [29].

– The silicon heterojunction bottom cell structure

The silicon heterojunction solar cell with interdigitated back contact has demonstrated better performance than the standard one. Using this structure in tandem configuration may improve its performance more.

– **Interface optimization**

The ITO/SnO_2 junction is used as a recombination layer for electrical contact between the sub-cells. However, ITO acts as an intermediate reflector, significantly reducing the light that is transmitted to the bottom silicon solar cell [28]. By reducing the thickness of this layer or by using recombination tunnel junctions, made of high reflection index material like hydrogenated microcrystalline silicon, current generation in the silicon heterojunction is enhanced leading to an improvement of the efficiency of the tandem solar cell. In fact, the tunnel junction is used between the perovskite solar cell and the silicon sub-cell in order to facilitate the passage of the electrons from the electron transport electrode for the perovskite solar cell to the p-type emitter of the silicon solar cell. This approach is widely used in tandem solar cells based on perovskite and micromorph silicon or copper indium gallium di-selenide [101].

Furthermore, the thickness of the recombination layer used in the tandem solar cells influences the resulting performance, as demonstrated in Figure 72 [29].

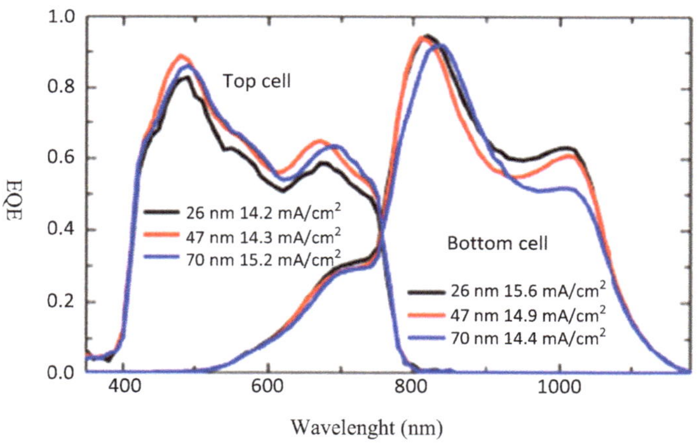

Figure 72: Effect of the thickness of IZO on the performances of the top cell (on the left) and those on the bottom one (on the right), with a 200 nm thickness of the spiro-OMeTAD.
Source: Reprinted with permission from WERNER, Jérémie, WENG, Ching-Hsun, WALTER, Arnaud, et al. Efficient monolithic perovskite/silicon tandem solar cell with cell area> 1 cm^2. The journal of physical chemistry letters, 2016, vol. 7, no 1, p. 161-166. Copyright 2016, American Chemical Society.

Figure 72 shows that as the thickness of the IZO layer increases, the current of the bottom cell decreases, unlike the one on the top cell [29]. Thus, for a thickness of 26 nm, the perovskite solar cell is the current limiting of the tandem solar cell unlike for the 70 nm thick IZO. The optimal situation is that where the currents of the two sub-cells are close and that corresponds in this case to thicknesses between 40 and 50 nm of IZO. Similar results were obtained using ITO as a recombination layer [29].

– Addition of an AR layer

The AR layer makes it possible to increase the tandem solar cell current, thanks to the reduction of the reflection losses. Indeed, the current of the aperture area of 1.22 cm^2 of the perovskite top cell was increased by 10% and 16% for the bottom cell [29]. Therefore, the tandem solar cell becomes limited by the top cell and not by the bottom one.

6 Conclusion

Perovskite/silicon tandem solar cell is considered as one of the best ways to improve the silicon solar cells, already matured. This tandem configuration is recent, but it is in continuous progress since it becomes attractive to scientific researchers and also for many industrials. Today, the perovskite/silicon tandem solar cell achieved a record efficiency of 29.1%. Moreover, a matrix of parameter remains to be optimized to produce perovskite/silicon tandem solar cells with efficiencies close to their theoretical limit of 46% [98], at a reasonable cost, which makes it possible to maintain a competitive advantage against the most efficient solar cells, but which are technologically the most expensive.

General conclusion

The recent perovskite solar cells are increasingly becoming competitive with silicon technology in terms of cost and efficiency. They offer multiple choices of materials, structure, and simple processing methods, leading to interesting properties compatible to be used in different solar cell applications either in simple or in tandem configuration. Furthermore, the stability issue of this technology becomes reduced as research progresses.

Perovskite-based solar cells show great potential to become the cheapest and most important source of solar energy in the near future. Currently, they achieve a record efficiency of 25.2%, rising to 29.1% when integrated on top of silicon-based solar cells.

Furthermore, the tandem solar cells containing the perovskite sub-cell are recent. They have already achieved record efficiencies higher than the current efficiency of the standalone solar cells. They are an attractive alternative to other conventional technologies and a way to improve their performances. However, several means of improvement remain to be investigated to approach the theoretical limit, while maintaining a competitive advantage over the most efficient solar cells, but which are technologically the most expensive.

In addition to researchers, industrialists show a great interest to the perovskite technology. Indeed, several collaborations have been launched, and they plan to start the commercialization of their developed perovskite/silicon tandem solar cell in 2021. In addition to the reduction of energy costs, these solar cells will be used in the construction field like sunbreakers on facades and noise barriers.

Finally, the perovskite-based solar cell brings new promise and inspiration to solar cells and their applications.

https://doi.org/10.1515/9783110760613-006

References

[1] Photovoltaique.info, “Les technologies de cellules solaires photovoltaïques,” 2016. http://www.photovoltaique.info/Les-technologies-de-cellules.html.

[2] O. O. Ogbomo, E. H. Amalu, N. N. Ekere, and P. O. Olagbegi, “A review of photovoltaic module technologies for increased performance in tropical climate,” Renew. Sustain. Energy Rev., no. March, pp. 0–1, 2016, doi: 10.1016/j.rser.2016.11.109.

[3] S. De Wolf, B. Niesen, and C. Ballif, “Efficient monolithic perovskite/silicon tandem solar cell with cell area >1 cm^2,” pp. 3–8, 2016, doi: 10.1021/acs.jpclett.5b02686.

[4] N.-G. Park, “Methodologies for high efficiency perovskite solar cells,” vol. 3, p. 15, 2016, doi: 10.1186/s40580-016-0074-x.

[5] NREL, “Best Research-Cell Efficiencies,” 2017. https://www.nrel.gov/pv/assets/images/efficiency-chart.png (accessed Apr. 14, 2017).

[6] M. K. Nazeeruddin, H. Snaith, and G. Editors, “Methylammonium lead triiodide perovskite solar cells: A new paradigm in photovoltaics,” MRS Bull., vol. 40, pp. 641–645, 2017, doi: 10.1557/mrs.2015.169.

[7] Q. Chen, N. De Marco, Y. Michael, T. Song, C. Chen, and H. Zhao, “Under the spotlight: The organic – inorganic hybrid halide perovskite for optoelectronic applications,” vol. 10, p. 355, 2015.

[8] M. A. Green, A. Ho-Baillie, and H. J. Snaith, “The emergence of perovskite solar cells,” Nat. Photonics, vol. 8, no. 7, pp. 506–514, 2014, doi: 10.1038/nphoton.2014.134.

[9] A. K. Jena, A. Kulkarni, and T. Miyasaka, “Halide perovskite photovoltaics: background, status, and future prospects,” Chem. Rev., vol. 119, no. 5, pp. 3036–3103, 2019, doi: 10.1021/acs.chemrev.8b00539.

[10] L. K. Ono, E. J. Juarez-Perez, and Y. Qi, “Progress on perovskite materials and solar cells with mixed cations and halide anions,” ACS Appl. Mater. Interfaces, vol. 9, no. 36, pp. 30197–30246, 2017, doi: 10.1021/acsami.7b06001.

[11] Y. M. Yang, J. You, G. Li, and Y. Yang, “Integrated perovskite/bulk-heterojunction toward efficient solar cells,” Nonoletters, vol. 15, p. 662, 2014.

[12] J. Im, I. Jang, N. Pellet, and N. Park, “Growth of CH3NH3PbI3 cuboids with controlled size for high-efficiency perovskite solar cells,” Nat. Nanotechnol., no. August, pp. 1–6, 2014, doi: 10.1038/nnano.2014.181.

[13] N. J. Jeon, J. H. Noh, Y. C. Kim, W. S. Yang, S. Ryu, and S. Il Seok, “Solvent engineering for high-performance inorganic–organic hybrid perovskite solar cells,” Nat. Mater., vol. 13, no. 9, pp. 897–903, 2014, doi: 10.1038/nmat4014.

[14] S. Gharibzadeh et al., “Record open-circuit voltage wide-bandgap perovskite solar cells utilizing 2D/3D perovskite heterostructure,” Adv. Energy Mater., vol. 9, no. 21, p. 1803699, 2019, doi: 10.1002/aenm.201803699.

[15] Kojima, A., Teshima, K., Shirai, Y. and Miyasaka, T., “Organometal halide perovskites as visible-light sensitizers for photovoltaic cells,” J. Am. Chem. Soc., vol. 131, p. 6050, 2009.

[16] M. Grätzel, “The light and shade of perovskite solar cells,” Nat. Mater., vol. 13, no. 9, pp. 838–842, 2014, doi: 10.1038/nmat4065.

[17] T.-T. Bui and F. Goubard, “Matériaux de transport de trous à base de petites molécules organiques pour cellules photovoltaïques hybrides solides,” Matériaux Tech., vol. 101, no. 1, p. 102, 2013, doi: 10.1051/mattech/2013056.

[18] M. M. Lee, J. Teuscher, T. Miyasaka, T. N. Murakami, and H. J. Snaith, “Efficient hybrid solar cells based on meso-superstructured organometal halide perovskites,” Science Express, p. 1228604, 2012.

https://doi.org/10.1515/9783110760613-007

[19] J. Burschka et al., “Sequential deposition as a route to high-performance perovskite-sensitized solar cells,” Nature, vol. 499, no. 7458, pp. 316–319, 2013, doi: 10.1038/nature12340.

[20] W.-Q. Wu, D. Chen, R. A. Caruso, and Y.-B. Cheng, “Recent progress in hybrid perovskite solar cells based on n-type materials,” J. Mater. Chem. A, vol. 5, no. 21, pp. 10092–10109, 2017, doi: 10.1039/c7ta02376f.

[21] W. Chen et al., “A comparative study of planar and mesoporous perovskite solar cells with printable carbon electrodes,” J. Power Sources, vol. 412, pp. 118–124, 2019, doi: 10.1016/j.jpowsour.2018.11.031.

[22] A. Fakharuddin, F. Di Giacomo, I. Ahmed, Q. Wali, T. M. Brown, and R. Jose, “Role of morphology and crystallinity of nanorod and planar electron transport layers on the performance and long term durability of perovskite solar cells,” J. Power Sources, vol. 283, pp. 61–67, 2015, doi: 10.1016/j.jpowsour.2015.02.084.

[23] M. F. Mohamad Noh et al., “The architecture of the electron transport layer for a perovskite solar cell,” J. Mater. Chem. C, vol. 6, no. 4, pp. 682–712, 2018, doi: 10.1039/c7tc04649a.

[24] T.-B. Song et al., “Perovskite solar cells: film formation and properties,” J. Mater. Chem. A, vol. 3, no. 17, pp. 9032–9050, 2015, doi: 10.1039/c4ta05246c.

[25] J. P. Correa Baena et al., “Highly efficient planar perovskite solar cells through band alignment engineering,” Energy Environ. Sci., vol. 8, no. 10, pp. 2928–2934, 2015, doi: 10.1039/c5ee02608c.

[26] E. H. Anaraki et al., “Highly efficient and stable planar perovskite solar cells by solution-processed tin oxide,” Energy Environ. Sci., vol. 9, no. 10, pp. 3128–3134, 2016, doi: 10.1039/c6ee02390h.

[27] L. Kranz et al., “High-efficiency polycrystalline thin film tandem solar cells,” J. Phys. Chem. Lett., vol. 6, no. 14, pp. 2676–2681, 2015, doi: 10.1021/acs.jpclett.5b01108.

[28] S. Albrecht et al., “Monolithic perovskite/silicon-heterojunction tandem solar cells processed at low temperature,” Energy Environ. Sci., vol. 9, no. 1, pp. 81–88, 2016, doi: 10.1039/C5EE02965A.

[29] Werner, J., Weng, C.H., Walter, A., Fesquet, L., Seif, J.P., De Wolf, S., Niesen, B., and Ballif, C., “Efficient Monolithic Perovskite/Silicon Tandem Solar Cell with Cell Area >1 cm 2,” J. Phys. Chem. Lett., vol. 7, p. 161, 2016, doi: 10.1021/acs.jpclett.5b02686.

[30] D. Wang, M. Wright, N. K. Elumalai, and A. Uddin, “Solar energy materials & solar cells stability of perovskite solar cells,” Sol. Energy Mater. Sol. Cells, vol. 147, pp. 255–275, 2016, doi: 10.1016/j.solmat.2015.12.025.

[31] M. A. Kamarudin et al., “Suppression of charge carrier recombination in lead-free tin halide perovskite via Lewis base post-treatment,” J. Phys. Chem. Lett., vol. 10, no. 17, pp. 5277–5283, 2019, doi: 10.1021/acs.jpclett.9b02024.

[32] F. Zhang, Q. Huang, J. Song, S. Hayase, J. Qu, and Q. Shen, “A new strategy for increasing the efficiency of inverted perovskite solar cells to more than 21%: high-humidity induced self-passivation of perovskite films,” Sol. RRL, vol. 4, no. 9, p. 2070094, 2020, doi: 10.1002/solr.202070094.

[33] M. Saliba et al., “Cesium-containing triple cation perovskite solar cells: improved stability, reproducibility and high efficiency,” Energy Environ. Sci., vol. 9, no. 6, pp. 1989–1997, 2016, doi: 10.1039/c5ee03874j.

[34] C. Wehrenfennig, G. E. Eperon, M. B. Johnston, H. J. Snaith, and L. M. Herz, “High charge carrier mobilities and lifetimes in organolead trihalide perovskites,” Adv. Mater., vol. 26, p. 1584, 2014, doi: 10.1002/adma.201305172.

[35] H. Zhou et al., “Interface engineering of highly efficient perovskite solar cells,” Science (80-.)., vol. 345, no. 6196, pp. 542–546, 2014, doi: 10.1126/science.1254050.

[36] J. H. Noh, S. H. Im, J. H. Heo, T. N. Mandal, and S. Il Seok, "Chemical management for colorful, efficient, and stable inorganic–organic hybrid nanostructured solar cells," Nano Lett., vol. 13, no. 4, pp. 1764–1769, 2013, doi: 10.1021/nl400349b.
[37] Y. Zhang et al., "Achieving reproducible and high-efficiency >21% perovskite solar cells with a presynthesized FAPbI3 powder," ACS Energy Lett., vol. 5, no. 2, pp. 360–366, 2019, doi: 10.1021/acsenergylett.9b02348.
[38] J. Zhu et al., "Grain enlargement and defect passivation with melamine additives for high efficiency and stable CsPbBr3 perovskite solar cells," ChemSusChem, vol. 13, no. 7, pp. 1834–1843, 2020, doi: 10.1002/cssc.201903288.
[39] Y. Wang et al., "Thermodynamically stabilized β-CsPbI3–based perovskite solar cells with efficiencies >18%," Science (80-.)., vol. 365, no. 6453, pp. 591–595, 2019, doi: 10.1126/science.aav8680.
[40] J. Y. Kim, J.-W. Lee, H. S. Jung, H. Shin, and N.-G. Park, "High-efficiency perovskite solar cells," Chem. Rev., vol. 120, no. 15, pp. 7867–7918, 2020.
[41] S. Akin, N. Arora, S. M. Zakeeruddin, M. Grätzel, R. H. Friend, and M. I. Dar, "new strategies for defect passivation in high-efficiency perovskite solar cells," Adv. Energy Mater., vol. 10, no. 13, p. 1903090, 2020, doi: 10.1002/aenm.201903090.
[42] G. Niu, W. Li, J. Li, X. Liang, and L. Wang, "Enhancement of thermal stability for perovskite solar cells through cesium doping," RSC Adv., vol. 7, no. 28, pp. 17473–17479, 2017, doi: 10.1039/c6ra28501e.
[43] R. Fu, Y. Zhao, W. Zhou, Q. Li, Y. Zhao, and Q. Zhao, "Ultrahigh open-circuit voltage for high performance mixed-cation perovskite solar cells using acetate anions," J. Mater. Chem. A, vol. 6, no. 29, pp. 14387–14391, 2018, doi: 10.1039/c8ta04453h.
[44] D. Zhou, T. Zhou, Y. Tian, X. Zhu, and Y. Tu, "Perovskite-based solar cells: materials, methods, and future perspectives," J. Nanomater., vol. 2018, pp. 1–15, 2018, doi: 10.1155/2018/8148072.
[45] M. Saliba et al., "Incorporation of rubidium cations into perovskite solar cells improves photovoltaic performance," Science (80-.)., vol. 354, no. 6309, pp. 206–209, 2016, doi: 10.1126/science.aah5557.
[46] K. Mahmood, S. Sarwar, and M. T. Mehran, "Current status of electron transport layers in perovskite solar cells: materials and properties," RSC Adv., vol. 7, no. 28, pp. 17044–17062, 2017, doi: 10.1039/c7ra00002b.
[47] R. Salazar et al., "Use of anodic TiO2 nanotube layers as mesoporous scaffolds for fabricating CH3NH3PbI3 perovskite-based solid-state solar cells," ChemElectroChem, vol. 2, no. 6, pp. 824–828, 2015, doi: 10.1002/celc.201500031.
[48] P. Zhang et al., "Perovskite solar cells with ZnO electron-transporting materials," Adv. Mater., vol. 30, no. 3, p. 1703737, 2017, doi: 10.1002/adma.201703737.
[49] X. Zhao, H. Shen, Y. Zhang, X. Li, and X. Zhao, "Aluminum doped zinc oxide as highly stable electron collection layer for perovskite solar cells," ACS Appl. Mater. Interfaces Alum., vol. 8, p. 7826, 2016, doi: 10.1021/acsami.6b00520.
[50] N. G. Park, D. Y. Son, J. H. Im, and H. S Kim, "11% efficient perovskite solar cell based on ZnO," J. Phys. Chem. C, vol. 118, p. 16567, 2014, doi: 10.1021/jp412407j.
[51] A. A. Bart Roose, J. C. Baena, K. C. Gödel, M. Graetzel, A. Hagfeldt, and U. Steiner, "Mesoporous SnO2 electron selective contact enables UV-stable perovskite solar cells," Nano Energy, 2016.
[52] R. E. Beal et al., "Cesium lead halide perovskites with improved stability for tandem solar cells," J. Phys. Chem. Lett., pp. 2–7, 2016, doi: 10.1021/acs.jpclett.6b00002.
[53] X. Zhao and M. Wang, "Organic hole-transporting materials for efficient perovskite solar cells," Mater. Today Energy, vol. 7, pp. 208–220, 2018, doi: 10.1016/j.mtener.2017.09.011.

[54] P. Vivo, J. K. Salunke, and A. Priimagi, “Hole-transporting materials for printable perovskite solar cells,” Mater. (Basel, Switzerland), vol. 10, no. 9, p. 1087, Sep. 2017, doi: 10.3390/ma10091087.
[55] O. Almora, L. dice, V. Roca, and G. G. Belmonte, “Perovskite solar cells: a brief introduction and some remarks,” Authorea. Authorea, Inc., doi: 10.22541/au.158575974.47833392.
[56] J. Y. Wang, F. C. Hsu, J. Y. Huang, L. Wang, and Y.-F. Chen, “Bi-functional polymer nanocomposites as hole-transport layers for efficient light harvesting: application to perovskite solar cells,” ACS Appl. Mater. Interfaces, vol. 7, no. 50, pp. 27676–27684, 2015, doi: 10.1021/acsami.5b08157.
[57] F. Liu, Q. Li, and Z. Li, “Hole-transporting materials for perovskite solar cells,” Asian J. Org. Chem., vol. 7, no. 11, pp. 2182–2200, 2018.
[58] J. A. Christians, R. C. M. Fung, P. V Kamat, and J. Accepted, “An inorganic hole conductor for organo-lead halide perovskite solar cells. improved hole conductivity with copper iodide,” J. Am. Chem. Soc., vol. 136, p. 758, 2013.
[59] S. Seo et al., “An ultra-thin, un-doped NiO hole transporting layer of highly efficient (16.4%) organic–inorganic hybrid perovskite solar cells,” Nanoscale, vol. 8, no. 22, pp. 11403–11412, 2016, doi: 10.1039/c6nr01601d.
[60] S. Laalioui, K. B. Alaoui, H. A. Dads, K. El Assali, B. Ikken, and A. Outzourhit, “Progress in perovskite based solar cells: Scientific and engineering state of the art,” Reviews on Advanced Materials Science. 2020, doi: 10.1515/rams-2020-0017.
[61] K. Wang et al., “p-type mesoscopic nickel oxide/ organometallic perovskite heterojunction solar cells,” Sci. Repport, pp. 1–8, 2014, doi: 10.1038/srep04756.
[62] C. Chiang and C. Wu, “Bulk heterojunction perovskite-PCBM solar cells with high fill factor,” Nat. Photonics, no. February, pp. 1–6, 2016, doi: 10.1038/nphoton.2016.3.
[63] P. Qin et al., “Inorganic hole conductor-based lead halide perovskite solar cells with 12.4% conversion efficiency,” Nat. Commun., no. May, pp. 1–6, 2014, doi: 10.1038/ncomms4834.
[64] M. K. Assadi, S. Bakhoda, R. Saidur, and H. Hanaei, “Recent progress in perovskite solar cells,” Renew. Sustain. Energy Rev., vol. 81, pp. 2812–2822, 2018, doi: 10.1016/j.rser.2017.06.088.
[65] N. Elumalai, M. Mahmud, D. Wang, and A. Uddin, “Perovskite solar cells: progress and advancements,” Energies, vol. 9, no. 11, p. 861, 2016, doi: 10.3390/en9110861.
[66] F. Giustino and H. J. Snaith, “Toward lead-free perovskite solar cells,” ACS Energy Lett., vol. 1, no. 6, pp. 1233–1240, 2016, doi: 10.1021/acsenergylett.6b00499.
[67] “La Pérovskite: une alternative aux cellules photovoltaïques en silicium,” 2017. https://www.lenergieenquestions.fr/la-perovskite-une-alternative-aux-cellules-photovoltaiques-en-silicium.
[68] W. Nie et al., “High-efficiency solution-processed perovskite solar cells with millimeter-scale grains,” Science (80-.)., vol. 347, no. 6221, pp. 522–525, 2015.
[69] L. K. Ono, M. R. Leyden, S. Wang, and Y. Qi, “Organometal halide perovskite thin films and solar cells by vapor deposition,” J. Mater. Chem. A, vol. 4, p. 6693, 2015, doi: 10.1039/C5TA08963H.
[70] S. Razza, S. Castro-Hermosa, A. Di Carlo, and T. M. Brown, “Research Update: Large-area deposition, coating, printing, and processing techniques for the upscaling of perovskite solar cell technology,” APL Mater., vol. 4, no. 9, p. 91508, 2016, doi: 10.1063/1.4962478.
[71] X. P. G. Qian Zhao, G. R. Li, J. Song, Y. Zhao, and Y. Qiang, “Improving the photovoltaic performance of perovskite solar cells with acetate,” Sci. Reports 6, Artic. number 38670, vol. 6, p. 38670, 2016.

[72] B. Li, J. Tian, L. Guo, C. Fei, T. Shen, X. Qu, G. Cao, and ACS, "Dynamic growth of pinhole-free conformal CH 3 NH 3 PbI 3 film for perovskite solar cells," ACS Appl. Mater. Interfaces Dyn., vol. 8, p. 4684, 2016, doi: 10.1021/acsami.5b11679.

[73] W. Nie et al., "Light-activated photocurrent degradation and self-healing in perovskite solar cells," Nat. Commun., vol. 7, no. May, pp. 1–9, 2016, doi: 10.1038/ncomms11574.

[74] A. Bera et al., "Fast crystallization and improved stability of perovskite solar cells with Zn2SnO4 electron transporting layer: interface matters," ACS Appl. Mater. Interfaces, vol. 7, no. 51, pp. 28404–28411, 2015, doi: 10.1021/acsami.5b09182.

[75] S. Ito, S. Tanaka, K. Manabe, and H. Nishino, "Effects of surface blocking layer of Sb2S3 on Nanocrystalline TiO2 for CH3NH3PbI3 perovskite solar cells," J. Phys. Chem. C, vol. 118, no. 30, pp. 16995–17000, 2014, doi: 10.1021/jp500449z.

[76] J. Cui et al., "Phosphor coated NiO-based planar inverted organometallic halide perovskite solar cells with enhanced efficiency and stability Jin," Appl. Phys. Lett. 109, vol. 171103, pp. 1–6, 2016, doi: 10.1063/1.4965838.

[77] D. Wang, M. Wright, N. K. Elumalai, and A. Uddin, "Stability of perovskite solar cells," Sol. Energy Mater. Sol. Cells, vol. 147, pp. 255–275, 2016, doi: 10.1016/j.solmat.2015.12.025.

[78] Y. Hu et al., "Hybrid perovskite/perovskite heterojunction solar cells," ACS Nano, vol. 10, no. 6, pp. 5999–6007, 2016, doi: 10.1021/acsnano.6b01535.

[79] N. Marinova, S. Valero, and J. L. Delgado, "Organic and perovskite solar cells: Working principles, materials and interfaces," J. Colloid Interface Sci., vol. 488, pp. 373–389, 2017, doi: 10.1016/j.jcis.2016.11.021.

[80] C. D. Bailie et al., "Semi-transparent perovskite solar cells for tandems with silicon and CIGS," Energy Environ. Sci., vol. 8, no. 3, pp. 956–963, 2015, doi: 10.1039/C4EE03322A.

[81] P. Löper et al., "Organic–inorganic halide perovskite/crystalline silicon four-terminal tandem solar cells," Phys. Chem. Chem. Phys., vol. 17, no. 3, pp. 1619–1629, 2015, doi: 10.1039/C4CP03788J.

[82] G. E. Eperon et al., "Perovskite-perovskite tandem photovoltaics with optimized band gaps," Science (80-.)., vol. 354, no. 6314, pp. 861–865, 2016, doi: 10.1126/science.aaf9717.

[83] Z. Fang et al., "Perovskite-based tandem solar cells," Sci. Bull., 2020, doi: 10.1016/j.scib.2020.11.006.

[84] R. Sheng et al., "Monolithic Wide Band Gap Perovskite/Perovskite Tandem Solar Cells with Organic Recombination Layers," J. Phys. Chem. C, vol. 121, no. 49, pp. 27256–27262, 2017, doi: 10.1021/acs.jpcc.7b05517.

[85] Z. Ma and B. He, "TCO-Si based heterojunction photovoltaic devices, solar cells – thin-film technologies," Intechopen.Com, p. 456, 2011, [Online]. Available: http://www.intechopen.com/source/pdfs/22803/InTech-Tco_si_based_heterojunction_photovoltaic_devices.pdf.

[86] B. P. Rand, J. Genoe, P. Heremans, and J. Poortmans, "Solar cells utilizing small molecular weight organic semiconductors," Prog. Photovolt Res. Appl., vol. 15, no. February 2013, pp. 659–676, 2015, doi: 10.1002/pip.

[87] A. Chirilă et al., "Potassium-induced surface modification of Cu(In,Ga)Se2 thin films for high-efficiency solar cells," Nat. Mater., vol. 12, no. 12, pp. 1107–1111, 2013, doi: 10.1038/nmat3789.

[88] IMEC, "Perovskite/CIGS tandem cell with Record Efficiency of 24.6 percent Paves the Way for Flexible Solar Cells and High-Efficiency Building-Integrated PV," 2018. https://www.imec-int.com/en/articles/perovskite-cigs-tandem-cell-with-record-efficiency-of-24-6-percent.

[89] P. Löper et al., "Organic–inorganic halide perovskite/crystalline silicon four-terminal tandem solar cells," Phys. Chem. Chem. Phys., vol. 17, no. 3, pp. 1619–1629, 2015, doi: 10.1039/C4CP03788J.

[90] K. Masuko et al., "Achievement of more than 25% conversion efficiency with crystalline silicon heterojunction solar cell," IEEE J. Photovoltaics, vol. 4, no. 6, pp. 1433–1435, 2014, doi: 10.1109/JPHOTOV.2014.2352151.

[91] M. Taguchi et al., "24.7% Record efficiency HIT solar cell on thin silicon wafer," IEEE J. Photovoltaics, vol. 4, no. 1, pp. 96–99, 2014, doi: 10.1109/JPHOTOV.2013.2282737.

[92] K. Yoshikawa et al., "Silicon heterojunction solar cell with interdigitated back contacts for a photoconversion efficiency over 26%," Nat. Energy, vol. 2, no. 5, p. 17032, 2017, doi: 10.1038/nenergy.2017.32.

[93] "best-research-cell-efficiencies.20200803.pdf." 2020, [Online]. Available: https://www.nrel.gov/pv/assets/images/efficiency-chart.png.

[94] K. A. Bush et al., "23.6% efficient monolithic perovskite/silicon tandem solar cells with improved stability Kevin," Nat. Energy, no. February, pp. 1–7, 2017.

[95] B. Chen et al., "Grain engineering for perovskite/silicon monolithic tandem solar cells with efficiency of 25.4%," Joule, vol. 3, no. 1, pp. 177–190, 2019, doi: 10.1016/j.joule.2018.10.003.

[96] J. Xu et al., "Triple-halide wide-band gap perovskites with suppressed phase segregation for efficient tandems," Science (80-.)., vol. 367, no. 6482, pp. 1097–1104, 2020, doi: 10.1126/science.aaz4639.

[97] P. V Oxford, "Oxford PV perovskite solar cell achieves 28% efficiency, 2018." 2019, [Online]. Available: https://www.oxfordpv.com/news/oxford-pv-perovskite-solar-cell-achieves-28-efficiency.

[98] W. Shockley and H. J. Queisser, "'Detailed balance limit of efficiency of p-n junction solar cells,'" J. Appl. Phys., vol. 32, pp. 510–519, 1961.

[99] F. Lang et al., "Perovskite solar cells with large-area CVD-graphene for tandem solar cells," J. Phys. Chem. Lett., vol. 6, no. 14, pp. 2745–2750, 2015, doi: 10.1021/acs.jpclett.5b01177.

[100] Z. Yin et al., "Graphene-Based materials for solar cell applications," Adv. Energy Mater., vol. 4, no. 1, pp. 1–19, 2014, doi: 10.1002/aenm.201300574.

[101] J. P. Mailoa et al., "A 2-terminal perovskite/silicon multijunction solar cell enabled by a silicon tunnel junction," Appl. Phys. Lett., vol. 106, no. 12, 2015, doi: 10.1063/1.4914179.

Index

2D (two-dimensional) 8
3D (three-dimensional) 2, 8

Ag (silver) 13, 33, 34, 49–51, 57–64
ALD (atomic layer deposition technique) 48, 49, 59
AR (antireflective) 6, 36, 37, 46, 57–59, 61, 68
a-Si: H (hydrogenated amorphous silicon) 44, 46, 55
Au (gold) 10, 11, 13, 37–39, 53, 54, 66
Au NPs (gold nanoparticles) 12

BCP (bathcuproin) 19, 33, 34
Br (bromine) 3, 7

C60 (fullerene) 11, 19, 33, 34, 63, 64
$CH_3NH_3PbI_3$ (methylammonium lead iodide) 15, 19, 36, 37, 41, 52
CIGS (copper indium gallium di-selenide) 1, 29, 32, 36, 37, 39–43, 67
CIS (copper indium di-selenide) 36
Cs (cesium) 3, 7, 8, 25, 63
CsI (cesium iodide) 59
c-Si (crystalline silicon) 17, 44–46, 49, 51, 60–64
c-TiO_2 (compact titanium dioxide) 36–38
CuSCN (copper thiocyanate) 13, 15
CVD (chemical vapor deposition) 59

DEPOT (poly(3,4-ethylenedioxythiophene)) 12, 33, 34
DIO (1,8-diiodooctane) 24
DMF (dimethylformamide) 17–19, 21, 59
DMSO (dimethylsulfoxide) 17, 21, 59

EA (ethylammonium) 8
EPFL (Federal Institute of Technology in Lausanne) 47
EQE (external quantum efficiency) 34, 35, 39, 41, 42, 54, 55, 66, 67
ETL (electron transporting layer) 5, 6, 9–11, 14, 15, 22, 33, 36

FA (formamidinium) 3, 7, 8, 25
FAI (formamidinium iodide) 59

FF (fill factor) 12, 35, 36, 38–40, 42, 51, 52, 54, 56–59, 62–64
FTO (Fluorine-doped Tin Oxide) 9, 36, 37

HIT (heterojunction solar cell) 44, 45, 47, 55
HTL (hole transporting layer) 11–15, 22, 33, 36

IO:H (hydrogenated indium oxide) 52–54
IPA (2-propanol) 19
ITO (indium tin oxide) 10, 13, 15, 32–34, 48–55, 57–64, 67
IZO (indium zinc oxide) 13, 52, 56, 57, 63, 67

J_{sc} (short-circuit current density) 3, 35, 36, 39, 40, 42, 43, 51, 52, 54, 56–58, 62–64
J–V (current density versus voltage measurements) 42, 43, 57, 58, 63

LIF (lithium fluoride) 6, 51, 58, 59, 60, 62, 64

MA (methylammonium) 3, 7, 8, 19
MAI (methyl ammonium iodide) 17, 19–21, 24, 53, 54
$MAPbI_3$ (methylammonium lead iodide) 3, 7, 10, 15, 18, 21, 36
$MAPbX_3$ (methylammonium lead halide) 7
$MASnX_3$ (methylammonium tin halide) 7
Mo (molybdenum) 39, 40, 41
MPP (maximum power point tracking) 54, 58

NiO (nickel oxide) 13, 14, 27, 33, 34, 58–60, 62, 64
NREL (National Renewable Energy Laboratory) 16, 36, 44, 45, 62, 64

P3HT (poly(3-hexylthiophene-2,5-diyl)) 12
P (pressure) 20, 40
Pb (lead) 3, 4, 7, 19
$PbBr_2$ (lead bromide) 7, 59
PbI_2 (lead iodide) 17, 19–21, 24, 53, 54, 59
PCBM (methyl[6,6]-phenyl-C61-butanoate) 11, 14, 19, 34, 52–55, 57–60, 62
PECVD (plasma-enhanced chemical vapor deposition) 46, 50, 55, 56, 61
PEIE (polyethylenimine) 52–54, 57
PEN (polyethylenenaphthalate) 13

https://doi.org/10.1515/9783110760613-008

PSCs (perovskite solar cells) 1, 2, 4–29, 32, 36–39, 43, 44, 47, 52, 54, 55, 66, 67, 69
PSS (poly(styrenesulfonate)12, 32poly(styrenesulfonate)) 11, 12, 14, 19, 33, 34
PTAA (polytriaryl amine) 12, 14, 15, 63

SEM (scanning electron microscope) 33, 37, 48, 52
Sn (tin) 3, 7
SnO_2 (tin dioxide) 9–11, 26, 33, 34, 48, 49, 51, 58–60, 62, 63, 67
spiro-OMeTAD (2,2′7,7′-tetrakis (N,N-di-*p*-methoxyphenylamine)-9,9′-spirobifluorene) 10–15, 36–38, 41, 48, 49, 51–55, 57, 66–67
SPO (stabilized power output) 34, 35, 51, 52

T (temperature) 1, 5, 10, 12, 17, 18, 24, 25, 28, 32, 39, 40, 48
TCO (transparent conductor oxide) 5, 6, 10, 14, 21, 38, 42

UV (ultraviolet) 11, 24, 26–28, 41, 47, 53, 54, 59

V_{oc} (open-circuit voltage) 3, 5, 12, 13, 35–37, 39–42, 44, 51, 52, 54, 56–58, 62–64

ZnO:Al (aluminum-doped zinc oxide) 36, 37, 39, 40, 50
ZTO (zinc tin oxide) 58–60, 62

μc-Si : H (hydrogenated microcrystalline silicon) 67

Bei Fragen zur Produktsicherheit wenden Sie sich bitte an:
If you have any questions regarding product safety,
please contact:

Walter de Gruyter GmbH
Genthiner Straße 13
10785 Berlin
productsafety@degruyterbrill.com